L. 13

LTIMATE INTERCEPTOR”

F-106 Delta Dart

in detail & scale

Bert Kinzey

squadron/signal publications

This book is a product of Detail & Scale, Inc., which has sole responsibility for the contents and layout, except that all contributors are responsible for the security clearance and copyright release of all materials submitted. Published by Squadron/Signal Publications, 1115 Crowley Drive, Carrollton, Texas 75011.

CONTRIBUTORS:

George Cockle
Ray Leader
Bob Mason
Don Spering
Robert J. Mills
The U.S. Air Force
Doug Barbie
Lloyd Jones
Warren Munkasy
Al Lloyd
Charles R. King
Convair Division of G.D.
Wallace T. VanWinkle

Detail & Scale wishes to express a special thanks to LTC J.C. Hicks, LT Johnnie Ainsle, and SGT Brandon Williams of the Public Affairs Office at Tyndall AFB, Florida for arranging for many of the detail photos in this book to be taken. A special thanks is also extended to S/SGT Free, AMN Paul, and SRA Gower, all of whom are ground crew personnel for F-106s at Tyndall AFB.

A very special thanks is due Captain Kaye Downing. Her great love of the "Six" caused her to go to exceptional lengths to coordinate with many offices and people in order to gain access to certain F-106s for detailed photo work. She also contributed other materials to this publication, and thereby made it a better book than it would have been without her help.

Many photos in this book are credited to their contributors. Photos with no credit indicated were taken by the author.

ISBN 1-888974-26-5

Above, Front Cover: The pilot of this F-106A from the 49th FIS deploys the parabrake and holds the nose of his aircraft high to use the large delta wing to slow the aircraft as he touches down on the runway. (Flightleader)

Above Right and Right, Rear Cover: Cockpit details and colors in an F-106B are revealed in these two photographs. The top photo was taken in the front cockpit, while the bottom photograph shows the rear cockpit.

INTRODUCTION

This head-on view of the prototype F-106A shows the sleek cross-section of the aircraft, the large intakes, and the boundary layer wing fences originally installed on the aircraft. (General Dynamics via Barbier)

The F-106 Delta Dart was built in fewer numbers than any of the other "Century Series" fighters, but for many aviation enthusiasts it is a favorite and is something special when compared to other aircraft. Developed as a pure interceptor, its only reason for existing was to shoot down other aircraft, as it had no capability to attack ground targets. Carrying its weapons in an internal bay, it presented an unusually clean appearance, with externally carried fuel tanks being the only stores to hang from beneath the wings.

The longevity of the aircraft was remarkable. It entered operational service in 1959, and remains in active Air Force and Air National Guard squadrons in 1983. In 1984 it will reach the quarter-century mark of service as the last units are phased out and changed over to new aircraft. Serving only in U.S. colors, the F-106 has thus remained the primary interceptor for America's Aerospace Defense Command (now under TAC) for a longer period of time than any other aircraft, and it remains doubtful that any other aircraft will fill this interceptor role for such a long period of time. Indeed, the F-106 may be the last of the pure interceptors. The F-15 Eagle, to which two F-106 squadrons recently converted, is not a pure interceptor, but falls more into the air superiority (dogfighter) or even the fighter-bomber catagories. Considering the politics and money constraints involved, it seems unlikely that another single-mission interceptor aircraft will be developed. Therefore, the F-106 seems destined to remain not only the best, but also the last of the line of jet interceptors that included the F-86D Saber, F-89 Scorpion, F-94 Starfire, F-101B Voodoo, and the F-102 Delta Dagger.

On the pages that follow is the most extensive detail coverage ever presented on the F-106. Every major change ever made to the aircraft is covered. These include the change to the main wheels, the new canopy, the addition of an in-flight refueling capability, the new supersonic fuel tanks, the addition of the Vulcan cannon, and much more. Other sections include details of the landing gear and wells, radar and avionics, armament, cockpit coverage, including both the "round eye" and tape type instruments, and the J-75 engine. Five-view 1/72nd scale drawings are included as are tables giving technical and performance data on the aircraft. The differences between the F-106A and F-106B are shown in photographs and are discussed in the text. For modelers, our usual Modeler's Section is included showing the available kits and decals.

As the last F-106s are retired from service, it seems appropriate to take this detailed look at the aircraft that has served so long as America's guardian of the skies - the F-106 Delta Dart.

DEVELOPMENTAL HISTORY

The second F-106A, 56-452 is shown about to touch down after a flight. Note the original style wheels and the boundary layer fence on the wing that was used on early Delta Darts. (Convair)

As World War II ended, the dawn of the jet age ushered in an era of unbelievable technological advances. Top speeds of aircraft doubled, then doubled again. Airframe designs changed drastically as supersonic speeds were approached then passed. New radar and computers formed the components of fire control systems to guide the aircraft to its target in any weather, then fire the new guided missiles that were also beginning to make their appearance as armament for the new aircraft.

But another aspect of military airpower was also undergoing radical change. Tactics used in World War II were as obsolete as the aircraft that were used to carry them out. Many changes in tactics were brought about by the technological advances that were giving aircraft capabilities never before imagined, but other changes were caused by the changing political structure of the world and the shift in balance of military power. This shift was dramatically emphasized when the Soviet Union exploded its first nuclear bomb, and then began building a bomber force to deliver the stockpile of nuclear weapons that it began to accumulate at an alarming rate.

In World War II, strategic bombing took the form of mass bombing raids consisting of large numbers of bombers dropping scores of bombs on a given target. But with the advent of nuclear weapons, a single bomber could destroy a city, and a relatively small number of bombers could deal a devastating blow to an entire nation. As America developed the B-52, this awesome power was dramatized by the claim that one B-52 could carry more destructive force than all of the bombs dropped by all of the bombers in World War II. Civil Defense messages on radio and television told the American public that the Soviet Union had the "bomb" and the means to deliver it. This brought about considerable alarm, and resulted in a lot of concrete being poured into backyard bomb shelters. Americans, who had never suffered a single bombing raid on their homeland during World War II, now feared destruction from the air in a nuclear war with the Soviets.

Military planners in the United States were equally concerned about this threat, and a large percentage of the defense budget was dedicated to the air defense of the North American continent. Air defense missiles were deployed like the Nike Ajax and later the Nike Hercules and Hawk. The Distant Early Warning (DEW) line was built to detect bombers approaching over the North Pole.

As a part of this defense, the concept of the interceptor was born. Fighters were no longer just fight-

These two photographs compare the F-106 to its predecessor, the F-102 Delta Dagger. At left is an F-102, and the right photo is of the second F-106A. The canopy, nose, and wing areas are very similar. The early F-106s had the same boundary layer fences as the F-102. Major areas of difference include the intakes, vertical tail, and the area around the engine exhaust. (Convair)

The fourth F-106A shows some updates over the second aircraft shown on the previous page. A slot in the wing has replaced the boundary layer fences, and a small cut or vent is on the spine. However this is not the refueling receptacle that was later added. Yet to be added is the IR sensor housing ahead of the windscreen.(Convair)

ers. Some were designed as fighter-bombers, others were to be air-superiority fighters (the classic fighter), and now entered the fighter interceptor. These interceptors were designed with only one mission in mind, to intercept and shoot down those threatening bombers before they could deliver their dreadful payloads. The F-86D was developed with this mission in mind, as was the F-89 and F-94. Later, the F-101B and the F-104A were designed and deployed as interceptors. But defense planners saw the need for an "ultimate interceptor" with a fully automatic fire control system that worked in consort with elaborate ground systems.

On January 13, 1949, an Advanced Development Objective called for a specially designed interceptor that would be operational in 1954. The aircraft that this project was to produce became known as the "1954 Ultimate Interceptor." Competitive entries were submitted by the industry, and the Corvair design was selected in September 1951. The production decision was made two months later in November.

As 1951 ended and 1952 began, it became evident that the "Ultimate Interceptor," with all of the capabilities that the Air Force desired, was not going to be ready by 1954. However an advanced interceptor was needed as soon as possible since interceptors then in service were considered woefully inadequate. To solve this dilemma, the Air Force decided on a two-phase development of the aircraft. To get a new interceptor as soon as possible, an interim, less capable aircraft would be built in limited numbers. This aircraft would be designated the F-102A, and would be followed by the F-102B which was to be the designation of what had been known as the "Ultimate Interceptor." While the two aircraft would have the same airframes, the F-102B was to have a higher thrust J-67 engine, and it would have the sophisticated electronic control system being developed by Hughes Aircraft.

But things did not progress as planned. Problems with the F-102A's fuselage design, and other problems in a number of subsystems, delayed the development of the F-102A and consequently the F-102B. By the end of 1954, no F-102As had been produced, and the F-102A was no longer considered an interim aircraft. Wright was a year behind schedule with the J-67, and in early 1955, the Pratt and Whitney J-75 was substituted as the engine for the future F-102B.

Looking down on the first F-106A. The area rule fuselage is shown to good effect in this view as are the boundary layer fences. (Convair)

An early F-106 poses with the components of the Hughes MA-1 fire control system. This was a large, bulky, and heavy system when compared to the solid state systems of today. Note the prototype IR seeker fairing ahead of the windscreen on the aircraft. (Hughes)

By late 1955, problems with the F-102A were solved to the point where the Air Force increased the production order for the aircraft to 749. At the same time seventeen F-102Bs were ordered, which was a reversal of the original plan to order the F-102A in small numbers and the -B in more substantial quantities.

As F-102A production began, work continued to progress on the F-102B. The seventeen aircraft ordered in November 1955 were earmarked for testing, and on June 17, 1956, the designation of the aircraft was changed from F-102B to F-106. This was due to the increasing number of differences between the two aircraft. In September 1956, the Air Force specified that the F-106 was to be able to intercept and destroy hostile airborne targets at a radius of 375 nautical miles and up to 70,000 feet. The entire intercept mission was to be under the control of the ground environment and the MA-1 fire control system in the aircraft. But the projected availability date of the aircraft was now late 1958. Later, the ceiling requirement was reduced to 55,000 feet.

The first prototype of the F-106A made its maiden flight on December 26, 1956, with the second aircraft flying on February 26, 1957. These flights were flown by Convair, and the Air Force first flew the aircraft on April 29, 1957. The aircraft flew at Mach 1.9, and reached 57,000 feet. However, early testing revealed that the aircraft's top speed and acceleration were short of expectations. Part of the problem stemmed from airframe design, principally with the intake ducts, but there were also problems with the J-75-P-9 engines.

In early 1957 the F-106 program was in jeopardy. Its high costs brought it to the point of reappraisal when the Air Force was faced with a severe fund shortage. Since so many delays had been encountered, other solutions had been found to meet the needs for an interceptor. These included building the F-102 in larger numbers, and the development of the F-101B interceptor. Thus the F-106 program had lost a lot of the priority status it had originally enjoyed when it was begun in 1949 and was envisioned as the "Ultimate Interceptor."

The Air Defense Command (ADC) fought for the development of the F-106, and was opposed to dropping either the F-101B or the F-106 in order to save one aircraft at the expense of the other. ADC believed that both aircraft were complimentary and both were needed. Finally, ADC won its fight, and the F-106 program survived. However, there were considerable changes. The number of F-106s desired by ADC was one-thousand, but this was reduced to about one-third that number. Therefore all test aircraft were converted to operational status in order to obtain as many operational aircraft as possible.

The F-106 entered operational service in May 1959, with the 498th Fighter Interceptor Squadron at Geiger AFB, Washington. The squadron was considered operationally ready as of October of that year.

During the early operational life of the F-106, several shortcomings were noted, and programs were undertaken to provide solutions. Improvements, to include a better supersonic ejection seat, an infrared search-and-track system, and an upgrading of the MA-1 fire control system, were all intergrated into the aircraft. Additionally, the aircraft was given a thermal flash blindness protection hood, and improvements were also made to the now-installed -17 engine.

The F-106 featured a unique "Y"-shaped control column which not only controlled the aircraft, but the radar as well. At one point, plans called for the column to be located to the right side of the cockpit so as to afford the pilot unobstructed visibility of the Horizontal Situation Indicator (HSI). This was located below the instrument panel and low down in the center of the cockpit. However, the decision was made to return the column to the conventional center position.

Another unusual feature of the cockpit was the vision splitter. The flat glass of the windscreen often

Another in-flight view of 56452 showing the clean sleek lines of the Dart. ***(Convair)***

The first F-106A pops its "chute" upon landing. Note that the buzz number on this side of this particular aircraft is ahead of the U.S. AIR FORCE on the intake. ***(Convair)***

presented a glare off of the glass on the opposite side. To eliminate this problem, a vision splitter was placed vertically between the two pieces of glass at the highest point of the windscreen. This vision splitter was simply a thin flat panel painted black, and the pilot could easily see to either side of it with no glare. Behind the splitter, at the top of the apex of the windscreen, was an optical sight for aiming the Genie rocket. This sight was later removed from all aircraft.

Other early minor changes to the F-106 included making the retractable upper anti-collision beacon non-retractable, and the main gear wheels were changed. An arresting hook was added, and later data link antennas were added just to the left of the hook. These and other changes are shown in photos on the following pages.

By 1965, the F-106 had served as ADC's first line interceptor for six years, and the Air Force began a modernization of the F-106 fleet. A new TACAN, the first to use microelectronic circuits, was added, and was one-third the size and weight of the system it replaced. The installation of an in-flight refueling capability was authorized, and was installed on F-106A and -B aircraft. Supersonic fuel tanks replaced the old subsonic tanks, and carried more fuel, thus increasing the Dart's radius of action. Later, a clear canopy improved the pilot's visibility, and a gun was authorized for close-in attack against hostile aircraft after the F-106 demonstrated a remarkable dog-fighting ability.

In 1969, ten years after the F-106 entered opera-

Close-up of the wing slot which replaced the boundary layer wing fences on the F-106.

tional service, the Minimum Essential Improvement in System Reliability (MEISR) program was approved which again improved the systems in the aircraft. At that time, the Air Force had considered several advanced manned interceptor (AMI) studies ranging from the YF-12A to the F-14, but had not selected a replacement for the now aging Delta Dart. Until a replacement could be found, the only alternative was to keep the F-106 as up-to-date as possible. However all the improvements took time to implement, and it was not until 1972 that the clear canopy was installed on F-106As. The "Six Shooter" cannon and gunsight took even longer, and was installed only in a portion of the aircraft. Radar homing and warning gear, tested on one F-106B, was dropped from the list of improvements to be added to the F-106, as was the addition of formation light panels. But with no replacement in sight, the F-106 was recertified for a longer service life, increasing it from 4000 to 8000 hours, after undergoing a flight-and-fatigue test program.

An early major effort toward improving radar detection range was conducted in 1959 and identified as the F-106C/D program. In this effort a 40-inch diameter parabolic antenna (functionally similar to the existing 23-inch antenna) was installed in an F-106A aircraft. The installation was accommodated by replacing the existing 7 foot length radome with a 12 foot length radome of larger diameter at the antenna location. This resulted in a 5 foot extension of the nose, and a gloved fairing of the fuselage aft of the radome split line over a length of approximately 15 feet to a point just forward of the engine inlets. F-106A S/N 57-0239 was modified for aircraft performance testing, and F-106A S/N 57-0240 was modified for radar system testing. The aircraft performance test vehicle was flown ten times in the Spring of 1959 prior to program curtailment. The radar system test vehicle was not flown.

Though flight test time of the F-106C/D program was limited to just ten flights, the pertinent aircraft performance data was obtained. On flight four, a speed of Mach 2 was attained. The acceleration from Mach 1 to Mach 2 was as fast as the best performance obtained with an F-106A. Pilots who flew the modified aircraft agreed it was virtually impossible to determine from aircraft performance and handling qualities whether they were flying the long nose version or a standard F-106A.

The F-106X proposal to provide a look-down radar system and shoot-down armament system led to a detailed systems study identified as the F-106E/F. The objective of the F-106E/F study from a radar system standpoint was to evaluate candidate systems which could provide both increased detection range and clutter visibility. The antenna configurations evaluated were of three different types: a parabolic dish, two flat plate planar arrays, and an electronically scanned phased array.

The accommodation of these candidate systems required the manufacture of a new radome of about 9 foot length, a structural section approximately 4 feet in diameter and 4 feet in length, and new skins over a length of about 6 feet. The revised nose lines were faired into the existing structure, forward of the canted bulkhead, so as to cause no adverse effect on airflow ahead of the engine inlet ducts.

Use of an F-106 aircraft as an avionics test bed for the complete F-15 avionics system was studied and proposed in 1969. The proposed installation of equipment included the F-15 radar and F-15 radome. Structural modifications were thus similar to those derived in the F-106E/F study. The new nose structure, which was similar to the existing structure, con-

This F-106A-120-CO shows some early changes to the F-106. The production IR seeker housing is present as are the wing slots and vent on the spine. The aircraft still has the early style wheels. This aircraft was used for testing, and has some equipment attached to the leading edge of the wing at the root. (Convair)

The first F-106B approaches for a landing. Changes made to the -A were also made to the F-106B. The slot on the spine is not the refueling receptacle added later. (Convair)

A production F-106B shows the differences from the first -B shown above. Note the difference in the main wheels, the addition of the in-flight refueling receptacle, and the fairing for the IR seeker. Also note the slot in the wing which replaced the boundary layer fences. (Cockle)

sisted of aluminum alloy upper and lower closed section torque boxes comprised of longerons, frames, and skins. The upper and lower torque boxes were joined by a centerline vertical shear web. Large hinged doors provided compartment access as on the existing aircraft. New fuselage skins replaced existing skins back to the canted bulkhead. Frames and bulkheads were brought to the new contour by addition of an angle section riveted to the existing frame or by shimming.

When no new interceptor materialized, and no follow-on version of the F-106 was accepted, the F-106A and -B continued in front line service with ADC into the 1980's. But age catches up with all aircraft, and the F-106 appeared at the 1982 William Tell weapons meet in relatively few numbers. F-15s had then replaced the F-106 in two former ADC squadrons, all of which had since been reassigned to TAC. The F-106 has probably made its last appearance at William Tell in active Air Force units. The last active F-106 squadrons will soon retire the aircraft.

With the F-15 replacing the F-106, an era in military aviation is coming to an end. The F-106 is going out as the last of the pure interceptors. The F-15 was designed as an air superiority fighter, and has significant fighter-bomber capabilities designed into its systems. Funding restraints have eliminated the concept of a single-mission aircraft such as the interceptors, and thus the F-106 becomes the last of the breed.

The F-106's operational life was highlighted by several significant events, one of which was the deployment to Korea in the wake of the Pueblo crisis. But perhaps the highlight for the F-106 came early-on. On December 15, 1959, an F-106 set a world speed record of 1,525.695 mph. Piloting the aircraft was Colonel Joe Rogers. What makes this so remarkable is that this record still stands as the speed record for a single-engine aircraft over twenty-three years later! While the F-106 was built in fewer numbers than any other "Century Series" aircraft, and in far less numbers than originally planned, the Dart still became not only the last, but the "Ultimate Interceptor."

F-106 ADVANCED PROPOSALS

The F-106X was proposed as an alternate to the Lockheed YF-12 aircraft. It featured large rectangular intakes and canard wings. (Convair)

F-106C RADAR INSTALLATION

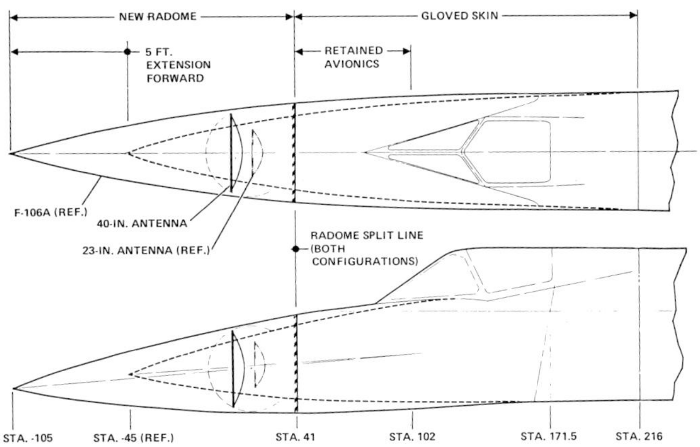

NEW RADOME
GLOVED SKIN
5 FT. EXTENSION FORWARD
RETAINED AVIONICS
F-106A (REF.)
40-IN. ANTENNA
23-IN. ANTENNA (REF.)
RADOME SPLIT LINE (BOTH CONFIGURATIONS)
STA. -105
STA. -45 (REF.)
STA. 41
STA. 102
STA. 171.5
STA. 216

F-106/F-15 AVIONICS TESTBED

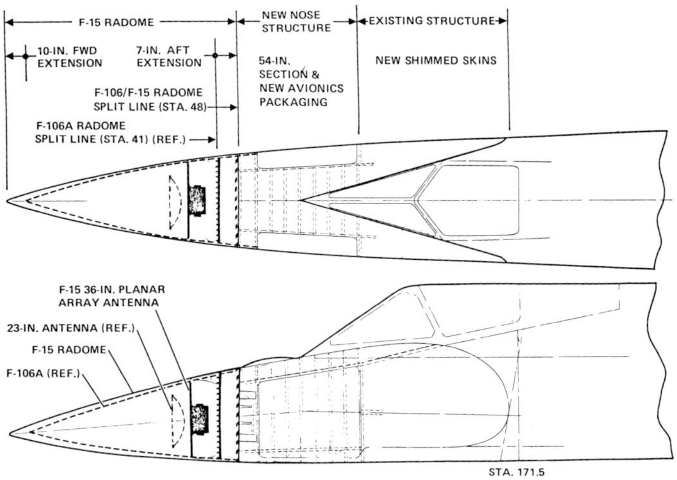

F-15 RADOME
NEW NOSE STRUCTURE
EXISTING STRUCTURE
10-IN. FWD EXTENSION
7-IN. AFT EXTENSION
54-IN. SECTION & NEW AVIONICS PACKAGING
NEW SHIMMED SKINS
F-106/F-15 RADOME SPLIT LINE (STA. 48)
F-106A RADOME SPLIT LINE (STA. 41) (REF.)
F-15 36-IN. PLANAR ARRAY ANTENNA
23-IN. ANTENNA (REF.)
F-15 RADOME
F-106A (REF.)
STA. 171.5

F-106E/F RADAR INSTALLATION STUDIES

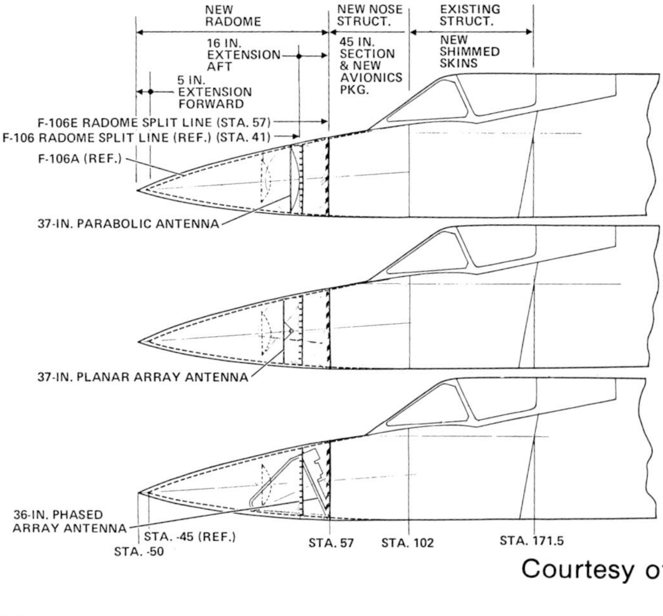

NEW RADOME
NEW NOSE STRUCT.
EXISTING STRUCT.
16 IN. EXTENSION AFT
45 IN. SECTION & NEW AVIONICS PKG.
NEW SHIMMED SKINS
5 IN. EXTENSION FORWARD
F-106E RADOME SPLIT LINE (STA. 57)
F-106 RADOME SPLIT LINE (REF.) (STA. 41)
F-106A (REF.)
37-IN. PARABOLIC ANTENNA
37-IN. PLANAR ARRAY ANTENNA
36-IN. PHASED ARRAY ANTENNA
STA. -45 (REF.)
STA. -50
STA. 57
STA. 102
STA. 171.5

CONFIGURATION COMPARISON

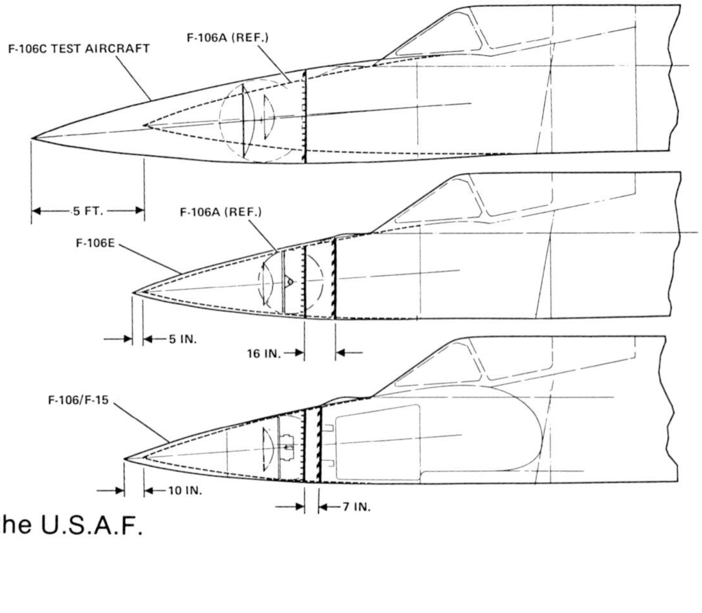

F-106C TEST AIRCRAFT
F-106A (REF.)
5 FT.
F-106A (REF.)
F-106E
5 IN.
16 IN.
F-106/F-15
10 IN.
7 IN.

Courtesy of the U.S.A.F.

F-106 UPDATES

FUEL TANKS

The F-106 was originally fitted with provisions for 230 gallon external fuel tanks as shown in the top photo. These tanks were subsonic. In June 1967, supersonic 360 gallon tanks were issued. These tanks were longer and more streamlined as shown in the lower photo. (Top Convair, Lower Spering)

ARRESTING HOOK

In May 1963, an arrestor hook was added to the "Six," although some test installations date back to the 1960-61 time frame. At left is a photo showing the original hook installation, while at the right are two photos of the later arrangement with a larger device to prevent the hook from catching on to something when retracted. In the two photos at the right, the data link antennas are also visible to the left of the hook. (Left photo, Barbier)

IR SEEKER

This photo shows the original prototype IR seeker installation on an early F-106A. (Convair)

Front view showing the seeker extended.

Right side view of the production IR seeker fairing. This installation was added in mid-February 1963.

Left side view of the production fairing with the seeker retracted.

OPTICAL SIGHT

The two photos above show the optical sight originally fitted for aiming the AIR-2 Genie rocket. This sight was removed after July 1972. (Barbier)

IN-FLIGHT REFUELING MODIFICATION

Originally the F-106 had a smooth spine with no slot or refueling receptacle. ***(Convair)***

This early F-106A had a small slot on the spine. This was replaced with an in-flight refueling receptacle beginning in early September 1967. ***(Convair)***

Looking forward into the vanes on the in-flight refueling receptacle on an F-106A. ***(Mason)***

The refueling receptacle on the F-106B is similar to the one on the F-106A, except that it is mounted at more of an angle and is not as rounded at the rear.

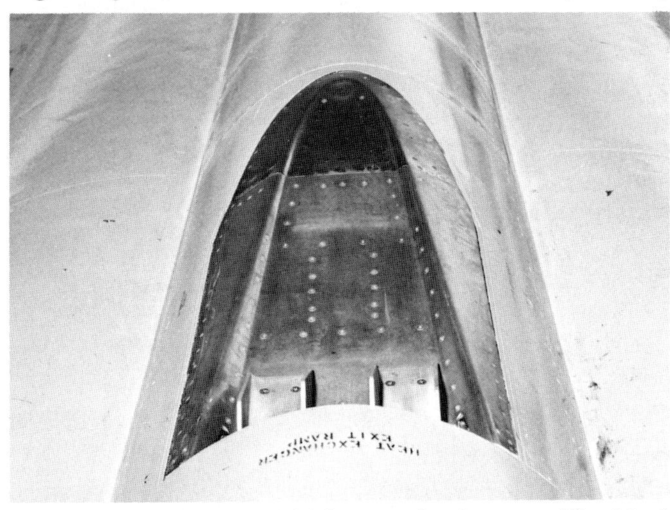

Looking aft at the F-106A receptacle. ***(Barbier)***

Close-up of the receptacle. Aft is to the right. During refueling, the aft portion hinges down into the aircraft allowing the probe from the tanker to be pushed down into the area immediately behind the bare metal slipway door.

With refueling receptacle open, F-106A, 90085, from the 48th FIS, moves in on a tanker for in-flight refueling on April 25, 1980. ***(Linn)***

MAIN GEAR WHEELS

The first F-106A shows the original spoked style main wheels in this early photo. ***(Convair)***

This production F-106A also shows the earlier style wheels as used up until the 1965-66 time frame. Also note the subsonic 230 gallon tanks. ***(Barbier)***

This close-up shows the later style wheels that were used beginning in 1965.

ROTATING BEACON

Prior to mid 1962, the upper fuselage rotating beacon was retractable. After that time it was fixed.

BUBBLE CANOPY

Prior to October 1972, the F-106A had a canopy with a bar across the top and two relatively flat side pieces of glass, as shown in these two views.

After October 1972, a new clear canopy was fitted that provided better all-around visibility. This was to help take better advantage of the F-106's dogfighting capability. The bar across the top was no longer present, and, as the photo at the lower left shows, there was a slight bubble shape to the canopy.

J-75 ENGINE

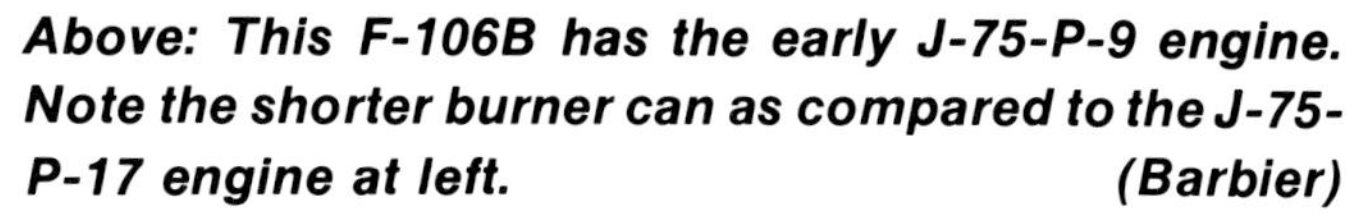

***Above:* This F-106B has the early J-75-P-9 engine. Note the shorter burner can as compared to the J-75-P-17 engine at left. (Barbier)**

***Left:* The standard production J-75-P-17 engine had a burner can that extended well beyond the fuselage skin. (Cockle)**

***Below left and right:* Two views of the J-75-P-17 removed from an F-106A. Note how big the afterburner can is compared to the engine itself.**

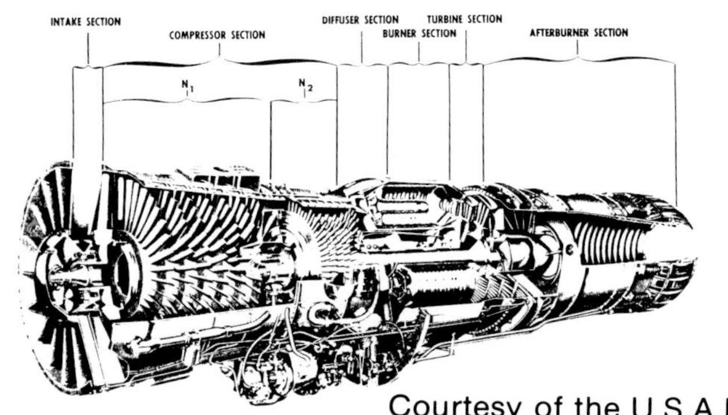

Courtesy of the U.S.A.F.

NOSE LANDING GEAR

Nose gear from the right. Note the light on the door and the actuating rod which opens and closes the door.

Nose gear from the right showing the wheel detail.

Close-up of the inside of the nose gear door.
(Spering)

Nose gear from behind and to the left.

Forward portion of the nose gear well.

Aft section of the nose gear well.

LEFT MAIN LANDING GEAR

Front view of the left main landing gear showing door braces and taxi light. (Spering)

Inside of left main gear wheel.

Looking out toward the left main landing gear from beneath the fuselage and slightly behind the gear.

Looking up into the left main gear well.

RIGHT MAIN LANDING GEAR

Right main gear from the inside front.

Right main gear well looking aft.

Right main gear well looking out toward the strut.

Right main gear from the inside and behind.

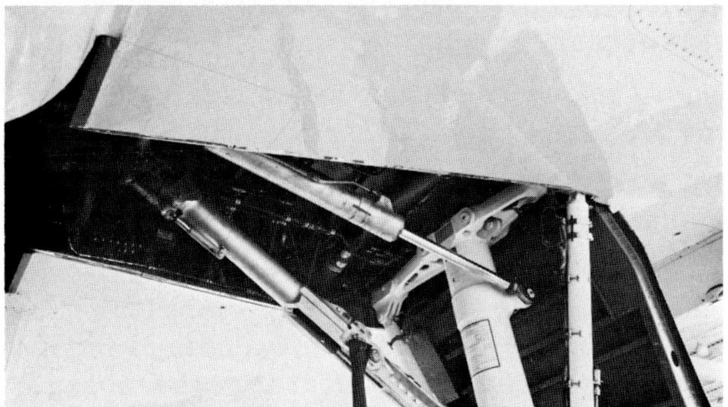

Right main gear well looking forward.

Outside of the right main landing gear showing the tapered door and later style wheel.

FUSELAGE GEAR WELL AND DOORS

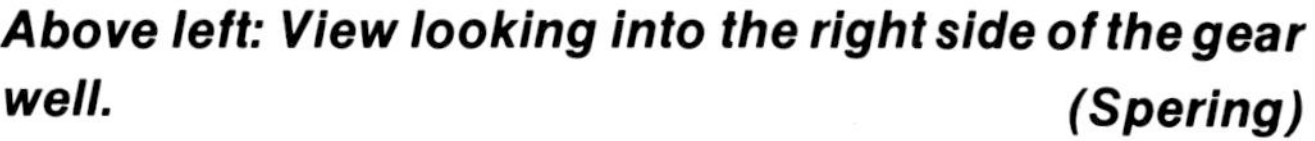

Above left: View looking into the right side of the gear well. *(Spering)*

Above right: Looking into the left side of the fuselage gear well. *(Spering)*

Right: Looking directly into the left side of the fuselage gear well. *(King)*

Below left: Fuselage gear doors showing rods and springs. These doors are usually down when the aircraft is on the ground. *(Spering)*

Below right: Doors from the front showing the angles of the rods and springs. The lower beacon and the emergency electrical generator can be seen ahead of the doors.

THE F-106A

Two F-106As from the 438th FIS. The near aircraft carries the older 230 gallon fuel tanks.(USAF via Barbier)

The F-106A was the primary version of the Dart, and a total of 277 were accepted by the Air Force. This total included the first two prototypes and the thirty-five first production aircraft which were originally used for testing. The flyaway cost per aircraft was 4.7 million dollars. Deliveries to operational squadrons began in 1959, and the last aircraft was delivered in December 1960.

During its operational life, the Delta Dart served in twenty-one active Air Force squadrons although some were redesignations of earlier units. Additionally, at least six Air National Guard units received the F-106.

The original YF-106As sported a bright orange and gray finish for high visibility, and had the same boundary layer wing fences that were used on the F-102. However these soon gave way to one small slot in each wing which performed the same function with less weight and drag. At first, the spine of the aircraft was straight with no breaks, but soon a small vent was added about a third of the way back from the canopy. Later, when the in-flight refueling receptacle was added, it was placed at the same point of this vent, and in fact, the vent remained at the front of the receptacle.

Most improvements made to the F-106A were also made to the F-106B. Those made only to the -A model included the addition of the gun (and not all -As were so fitted) and the addition of the clear canopy.

F-106A, 90122, of the 48th FIS as photographed on June 21, 1981. (Cockle)

F-106A TECHNICAL DATA

Mission and Description

Navy Equivalent: None   Mfr's Model: 8-24

The principal mission of the F-106A is the interception and destruction of attacking enemy aircraft and airborne missiles, having all weather and day or night characteristics.

This airplane incorporates a delta wing with a cambered leading edge extending from wing root to wing tip and swept tail surface. Control surfaces are power operated.

The fuel system is pressurized, air is bled from the engine compressor section and is used to pressurize the fuel tanks to reduce fuel evaporation and to provide for fuel transfer and CG control inflight.

The airplane has the capability for air refueling from flying boom equipped tankers.

Armament consists of four AIM missiles and an AIR-2 rocket or, on some airplanes, a gun system replaces the AIR-2 rocket. All armament is contained in a missile bay in the bottom portion of the fuselage. Firing of the armament is either manual or automatic. The components of the MA-1 Aircraft and Weapons Control System provides automatic radar searching and tracking, directs the airplane on a lead-collision attack and automatically fores the armament.

External fuel tanks can be added to increase range. The tanks can be refueled in flight and need not be jettisoned for combat since they do not restrict airplane speed or load factor when empty.

Development

Similar to the F-102A except for the J75 engine in lieu of the J57, redesigned tail, addition of fuselage fuel tanks, armament changes, and completely new electronic system.

Previously designated F-102B.

First Flight (Prototype)	Dec 56
First Acceptance	Oct 58
Production Status	Completed

WEIGHTS

Loading	Lb	L. F.
Empty	24,861(A)	
Basic	25,130(A)	
Design	34,522	7.0
Combat	*33,570	7.0
Max T.O.	† 41,831	5.0
Max Land	‡ 36,114	2.0

(A) Actual
* For basic mission
† Limited by space
‡ Limited by design

ELECTRONICS

Interceptor System, Aircraft and Weapons Control, Type MA-1 (Hughes Aircraft Corp.)
For detailed breakdown of MA-1 components, reference Convair Report ZM-8-452.

POWER PLANT

Nr & Model	(1) J75-P17
Mfr	Pratt & Whitney
Engine Spec Nr	A-2625
Type	Axial
Length	237.6
Diameter	44.25
Weight (dry)	5875 lb
Tail Pipe	Auto, Two-Position
Augmentation	Afterburning

ROCKETS

Nr	Type	Location
1	AIR-2A	Fuselage
	PLUS	
4	AIM-4F	Fuselage
	OR	
4	AIM-4G	Fuselage
	OR	
2	AIM-4F	Fuselage
2	AIM-4G	Fuselage

Courtesy of the U.S.A.F.

GUNS

No.	Type	Size	Rds ea	Location
*1	M61A1	20mm	625	Missile bay

* Installed in some aircraft; replaces AIR-2A rocket in center aft missile bay.

FUEL

Location	Nr Tanks	Gal
Fuselage	1	240
Wg. Internal	4	1034
Transfer	2	210
Transfer Lines		30
External Tanks	2	716
	Total	2230
Grade		JP-4
Specification		MIL-T-5624

OIL

Engine	1	(tot) 4.5
Specification		MIL-L-7808

ENGINE RATINGS

S.L. Static	LB	† RPM	MIN
Max:	*24,500	6440/8940	5
Mil:	16,100	6440/8940	30
Nor:	14,300	6080/8700	Cont

* With afterburner operating

† First figure represents the RPM of low pressure spool while the second is that of the high pressure spool.

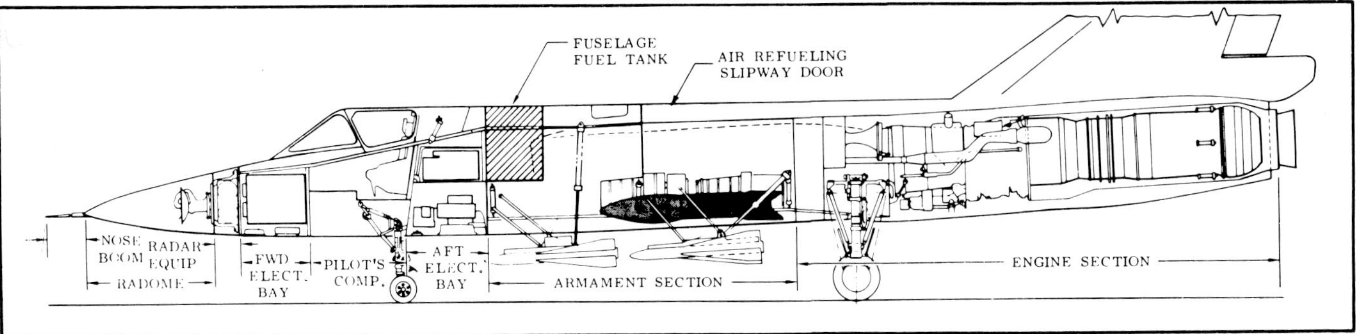

F-106A PERFORMANCE DATA

Loading and Performance — Typical Mission

CONDITIONS			MAXIMUM INTERNAL FUEL MISSION			EXTERNAL FUEL MISSION	
			POINT INTERCEPT I	AREA INTERCEPT II	FERRY RANGE III	AREA INTERCEPT IV	FERRY RANGE V
TAKEOFF WEIGHT		(lb)	36,663	36,663	35,240	41,831	40,408
Fuel at 6.5 lb/gal (grade JP-4)		(lb)	9841	9841	9841	14,495 (4)	14,495 (4)
Military load (missiles)	(5)	(lb)	594	594	—	594	—
Military load (rockets)	(6)	(lb)	829	829	—	829	—
Wing loading		(lb/sq ft)	52.8	52.8	50.7	60.2	58.1
Minimum speed (power off)	(9)	(kn)	150	150	150	150	150
Takeoff ground run	(1)	(ft)	3300	3300	3060	4100	3875
Takeoff to clear 50 ft	(1)	(ft)	5100	5100	4850	6000	5750
Rate of climb at SL		(ft/min)	39,000 (1) (8)	8850 (2) (8)	8950 (2)	7100 (2) (8)	7200 (2)
Time: SL to 40,000 ft	(8)	(min)	3.2 (1) (7)	7.9 (2)	7.5 (2)	12.8 (2)	11.5 (2)
Time: SL to 50,000 ft	(8)	(min)	6.0 (1) (7)	15.8 (2) (11)	15.6 (2) (11)	18.1 (2) (11)	17.7 (2) (11)
Service ceiling (100 ft/min)		(ft)	51,400 (1) (8)	44,200 (2) (8)	43,500 (2)	41,300 (2) (8)	40,200 (2)
COMBAT RANGE		(n mi)	—	—	1108	—	1718
COMBAT RADIUS		(n mi)	—	417	—	725	—
Average cruise speed		(kn)	—	516	516	516	516
Initial cruising altitude		(ft)	—	38,700	39,600	36,300	37,800
Final cruising altitude		(ft)	—	41,600	42,300	41,600	42,000
Total mission time		(hr)	—	1.70	2.15	2.89	3.65
TOTAL MISSION TIME	(7)	(hr)	1.87	—	—	—	—
Intercept altitude		(ft)	50,650	—	—	—	—
COMBAT WEIGHT		(lb)	33,570	31,770	26,987	33,760	27,728
Combat altitude		(ft)	50,650	51,600	42,300	50,600	42,000
Combat speed	(1)	(kn)	588	588	—	585	—
Combat climb	(1)	(ft/min)	500	500	9750	500	9750
Combat ceiling (500 ft/min)	(1)	(ft)	50,650	51,600	54,400	50,600	53,400
Service ceiling (100 ft/min)	(2)	(ft)	51,100 (1)	45,200	47,800	44,300	46,700
Maximum rate of climb at SL	(1)	(ft/min)	41,200	43,500	50,200	40,800	45,500
Maximum speed at 35,000 ft	(1) (10)	(kn)	1153	1153	1153	1153	1153
Basic speed at 50,000 ft	(1)	(kn)	1118	1135	1135	1116	1135
LANDING WEIGHT		(lb)	27,952	28,438	26,987	28,675	27,728
Ground roll at SL		(ft)	4200	4260	4090	4280	4250
Ground roll (auxiliary brake)	(12)	(ft)	2840	2920	2730	2940	2820
Total from 50 ft		(ft)	5620	5680	5510	5700	5590
Total from 50 ft (auxiliary brake)	(12)	(ft)	4300	4360	4170	4380	4250

NOTES

(1) Maximum thrust
(2) Military thrust
(3) Deleted
(4) With 716 gallons external fuel
(5) Four AIM-4F or 4G missiles
(6) One AIR-2A
(7) Includes time for take-off and acceleration to climb speed.
(8) Considers weight reduction due to fuel used.
(9) Onset of heavy buffet
(10) Design speed limit (M = 2.0)
(11) Time to service ceiling
(12) 14.5 ft (flat diameter drag chute plus speed brakes.

PERFORMANCE BASIS:
(a) Data source: Flight Test Service aircraft.
(b) Performance is based on powers shown on page 3.

F-106A WALK-AROUND

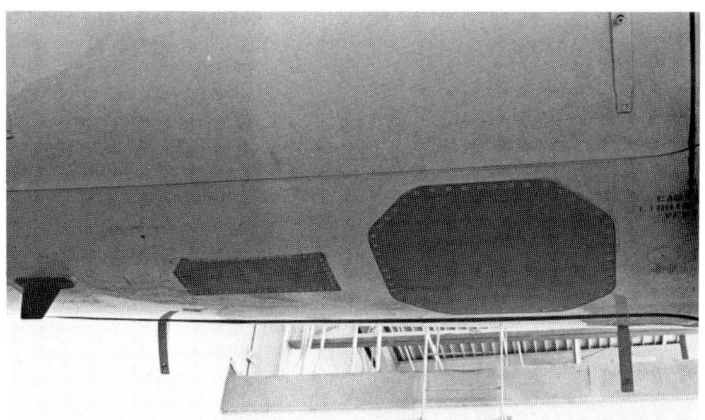

***Above left:* Head-on nose view of an F-106A.**

***Above right:* External electrical power connection located on the left side of the nose section.**

***Left:* Antennas located under the nose just ahead of the nose gear. The small blade antenna is the TACAN antenna, the strip antenna is the Marker Beacon antenna, and the large antenna is the AN/ARA-25 Direction Finder Antenna. Aft is to the right in this photo.**

***Below:* Raising and lowering mechanism for the canopy located just behind the seat. (Spering)**

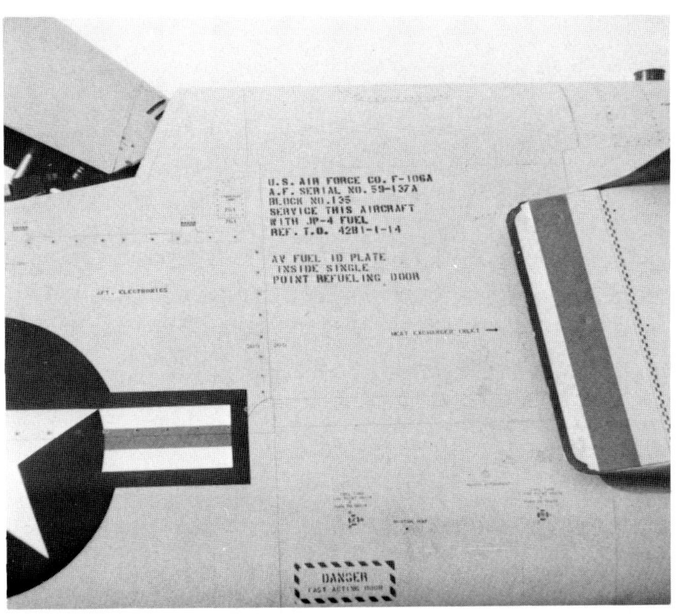

Above left: ***Typical data block stencil located on the left side of the fuselage just ahead of the intake.***

Above right: *Left intake detail.*

Right: ***Position light located on the left wing tip just outboard of the aileron. The light goes all the way through the wing. Note the usual droop to the aileron when the aircraft is on the ground.***

Below: Left side of the vertical tail. Note the probes on the leading edge of the tail, and also notice the position light just above the rudder. The black area on the leading edge and top of the tail is the UHF/TACAN antenna. Also clearly visible in this photo is the slot in the wing. ***(Cockle)***

***Above left:* Open speed brake from behind.(Cockle)**

***Above right:* Ram Air Turbine (RAT) located on the fuselage just in front of the main landing gear wells.**

***Left:* Data Link Antennas located just to the left of the tail hook.**

***Below left:* Open access door showing the hydraulic pump and accumulator for the weapons control system radar transmitter. This door is located on the right side of the aft fuselage behind the main gear wells.**

***Below right:* Position light on the underside of the right wing tip.**

Above: *Right side of center fuselage section showing the right intake and single point ground refueling receptacle. Note that the leading edge of the intake is bare metal.* ***(Barbier)***

Right: *Close-up of the single point ground refueling receptacle.*

Below: *This shot, taken from above, provides a good plan view of the F-106A. Note that the aircraft has the old style canopy.* ***(Spering)***

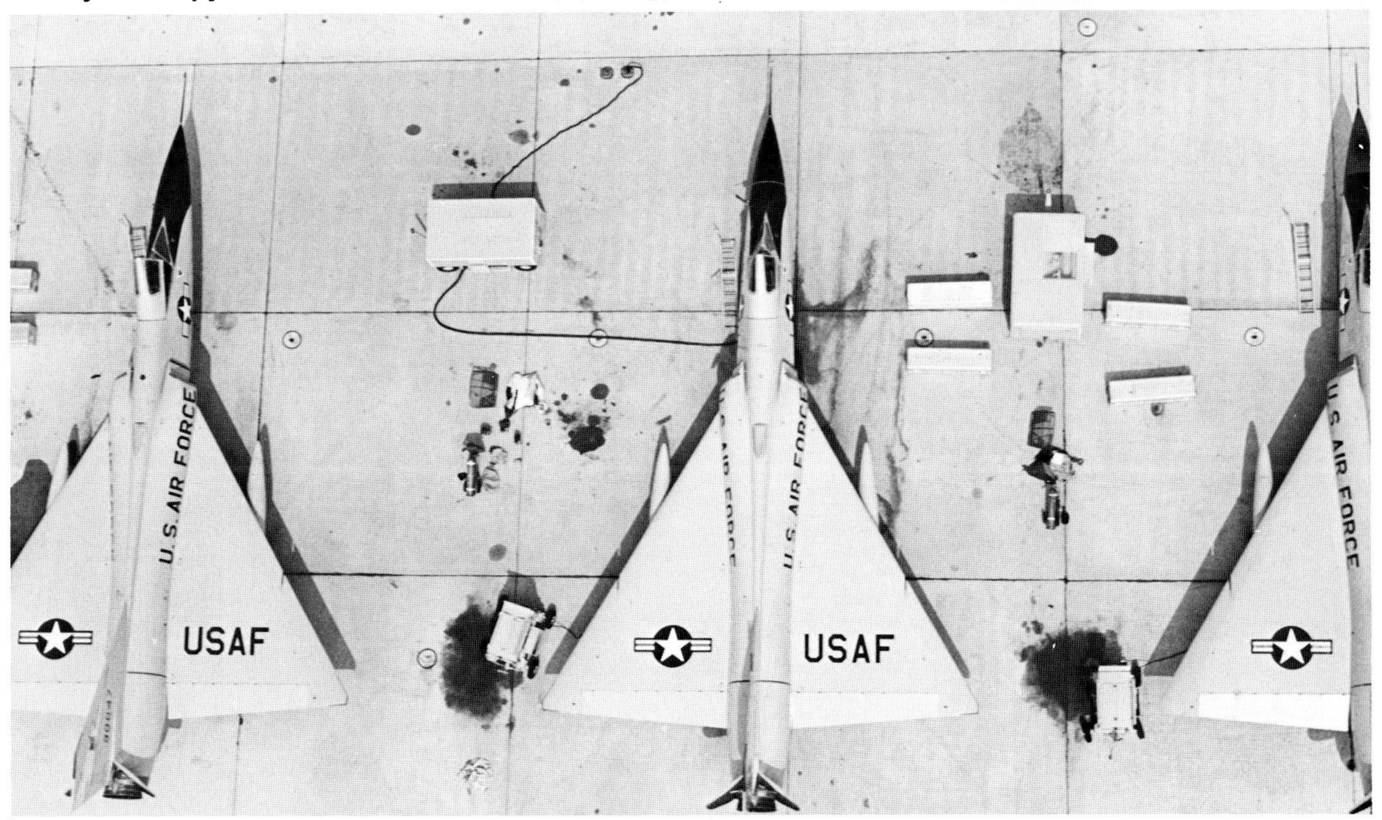

F-106A COCKPIT LAYOUT

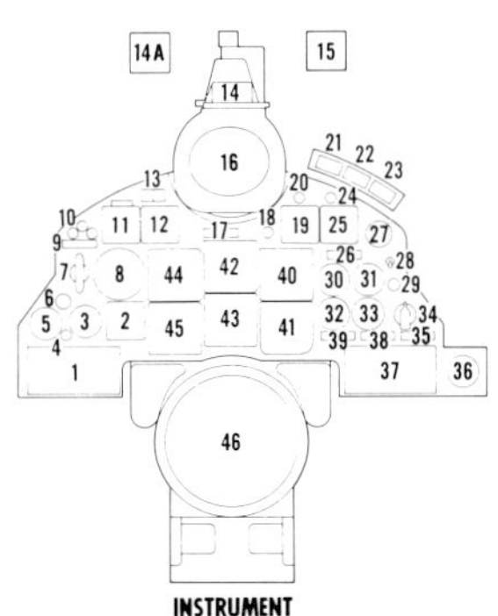
INSTRUMENT PANEL

F-106A cockpit -typical

CONVENTIONAL INSTRUMENT DISPLAY

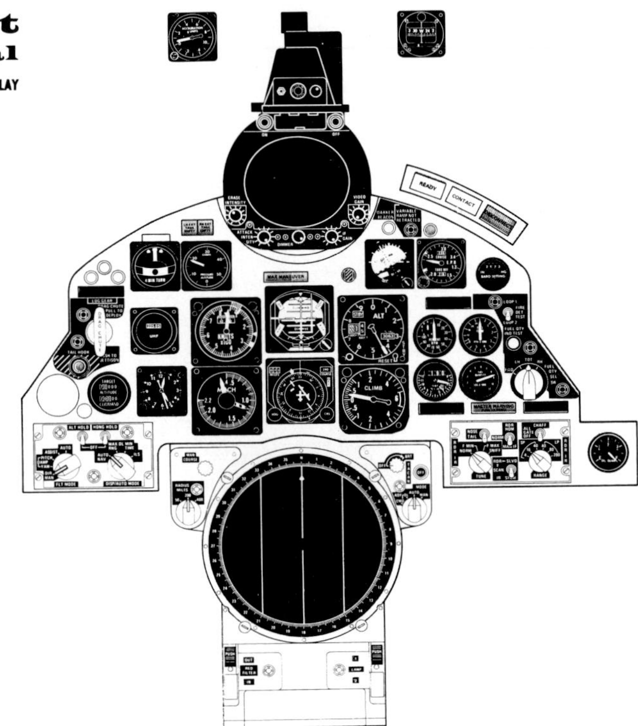

Courtesy of the U.S.A.F.

1. Flight Modes Panel
2. Clock
3. Command and Target Altitude Indicator
4. Blank
5. Blank
6. Tail Hook Down Button and Light
7. Drag Chute Handle
8. UHF Remote Indicator
9. Landing Gear Warning Light
10. Landing Gear Position Lights
11. Turn-and-Slip Indicator
12. Cabin Pressure Altitude Gage
13. External Tank Empty Lights
14. Radar Scope Recorder
14A. Accelerometer
15. Standby Compass
16. Radar Scope
17. Maximum Maneuver Warning Light
18. Computer Mode Indicator
19. Standby Attitude Indicator
20. Marker Beacon Light
21. Air Refueling Ready Light
22. Air Refueling Contact Light
23. Air Refueling Disconnect Light
24. Variable Ramp Warning Light
25. Engine Pressure Ratio Gage
26. Engine Fire Warning Light
27. Barometer Setting Control
28. Engine Fire Warning Test Switch
29. Fuel Quantity Gage Test Button
30. Tachometer
31. Fuel Flow Indicator
32. Exhaust Gas Temperature Gage
33. Fuel Quantity Gage
34. Fuel Quantity Gage Selector Switch
35. Hydraulic Pressure-Low Warning Light
36. Nucleonic Oil Quantity Indicator
37. Radar/IR Selector Panel
38. Master Warning Light
39. Canopy Unlocked Warning Light
40. Altimeter
41. Vertical Velocity Indicator
42. Attitude Director Indicator
43. Horizontal Situation Indicator
44. Airspeed-Angle of Attack Indicator
45. Mach Indicator
46. Tactical Situation Display (TSD)

LEFT CONSOLE

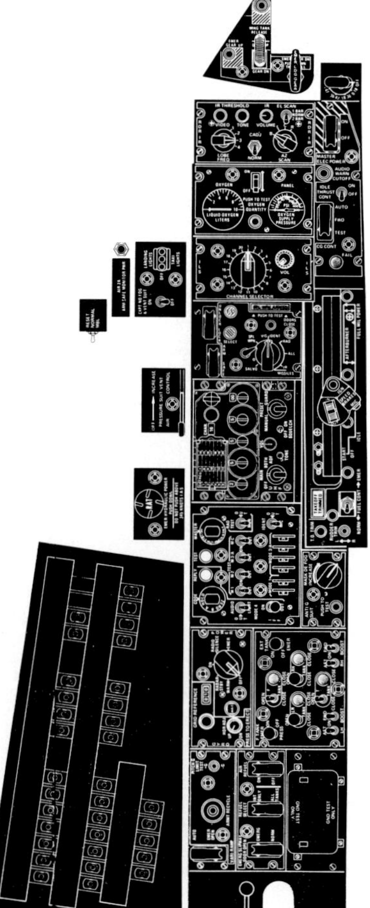

1. Landing Gear Emergency Up Button
2. Landing Gear Handle
3. External Wing Tanks Release Button
4. Landing Gear Emergency Extension Handle
5. Altitude Band Switch
6. Master Electrical Power Switch
7. Landing Gear Audio Warning Cutoff Button
8. Idle Thrust Control Switch
9. CG Control Switch
10. CG Transfer Test Failure Light
11. Radar/IR Control Panel
12. Oxygen Control Panel
13. ILS Channel Selector Panel
14. Armament Control Panel
15. Throttle Quadrant
16. Throttle
17. UHF Control Panel
18. Takeoff Trim Light
19. Fuel Control Switch
20. Rudder Trim Switch
21. IFF Control Panel
22. Mask Defog Rheostat
23. Anti-G Suit Test Button
24. MA-1 Power Control Panel
25. Fuel Control Panel
26. MA-1 Test Panel
27. Cabin Air Selector Handle
28. Air Refueling Panel
29. Variable Ramp Switch
30. Armament Recycle Button
31. Pitch G Limit Test Switch
32. Cockpit Left Fuse/Circuit Breaker Panel
33. Ram Air Turbine (RAT) Handle
34. Pressure Suit Control Handle
35. Reset/MBL Switch
36. Cockpit No-Fog and Ventilated Suit Switch
37. Landing and Taxi Light Switch
38. AIR-2A Arm/Safe/Monitor Power Circuit Breaker

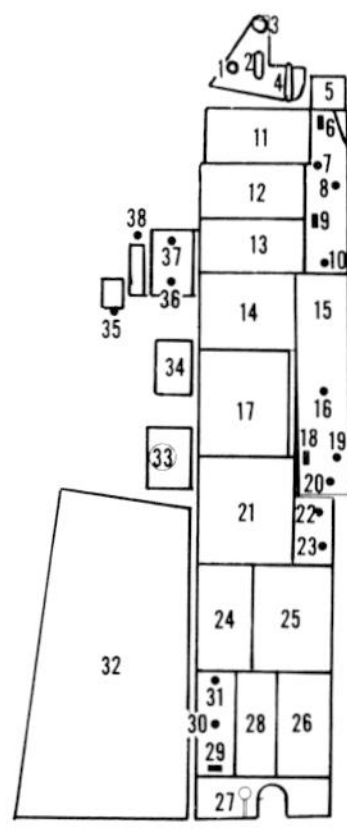

RIGHT CONSOLE

1. Master Warning Light Panel
2. Primary Hydraulic System Pressure Gage
3. Secondary Hydraulic System Pressure Gage
4. Oil Pressure Gage
5. IFF Caution Light
6. Generator Switch
7. No. 3 Fuel Tank Switch
8. Canopy Switch
9. Data Link Antenna Switch
10. Canopy Latch Handle
11. ATG Switch
12. Map Reading Light Switch
13. Warning Lights Test Button
14. Map Reading Light
15. EGT Spread Button
16. Emergency AC Generator Switch
17. Windshield Anti-Icing, Antifog Switches
18. Blank
19. TACAN-ADF Selector Switch
20. Engine Anti-Ice Warning Test Button
21. Rain Removal Switch
21A. Altitude Warning Selector
22. Ejection Seat Ground Safety Pin Stowage
23. Cabin Temperature Control Knob
24. Pitot Heat Switch
25. Canopy Antifog Switch
26. Surface and Engine Anti-Icing Switch
27. Compass System Controller
28. Cockpit Right Fuse/Circuit Breaker Panel
29. MATTS Switch
30. Map and Data Case
31. Lighting Control Powerstats
32. Lighting Control Panel
33. Data Link Control Panel
34. Auto-Navigation Homing Point Selector
35. TACAN Control Panel
36. Refrigeration Unit Switch
37. Cabin Air Selector Switch

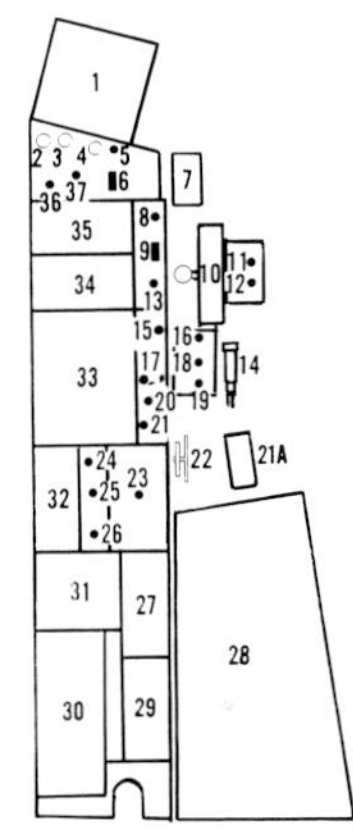

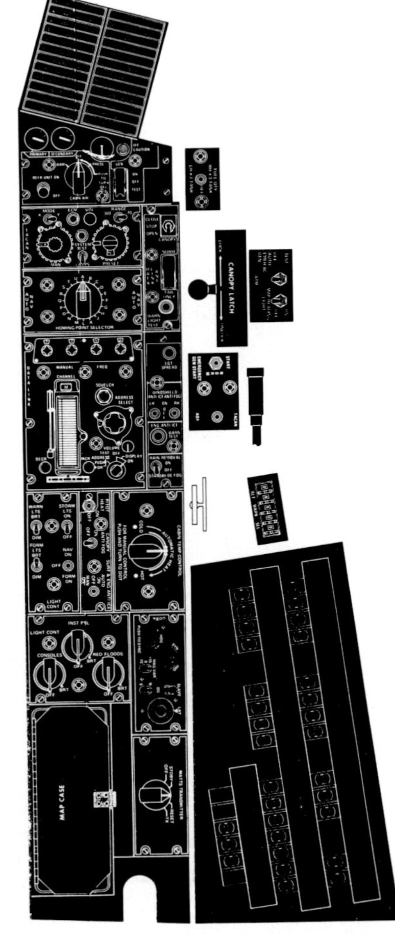

F-106A cockpit -typical

INTEGRATED FLIGHT INSTRUMENT SYSTEM

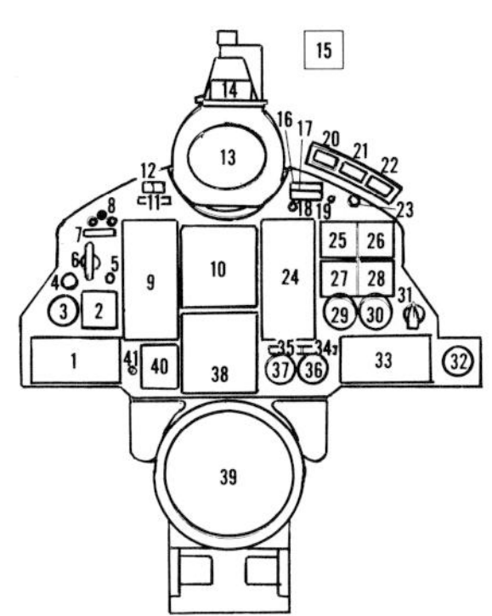

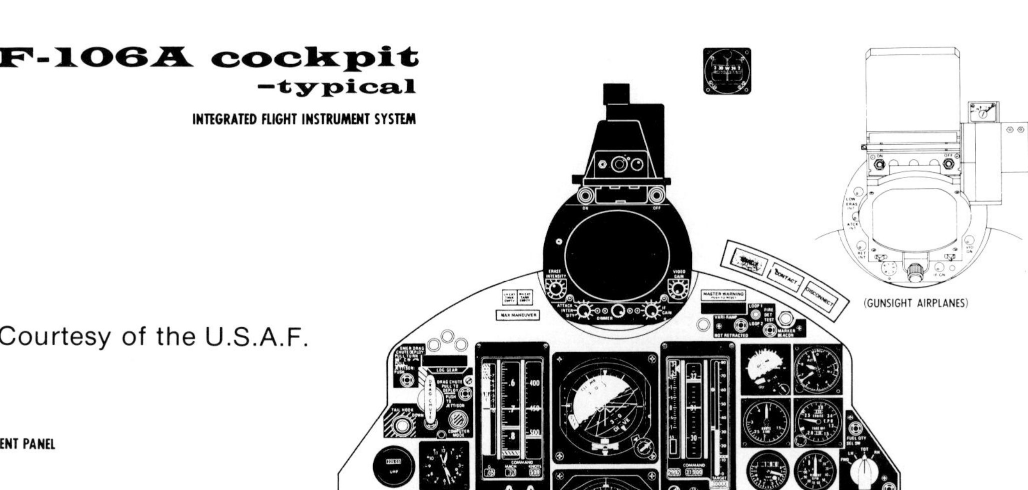

(GUNSIGHT AIRPLANES)

Courtesy of the U.S.A.F.

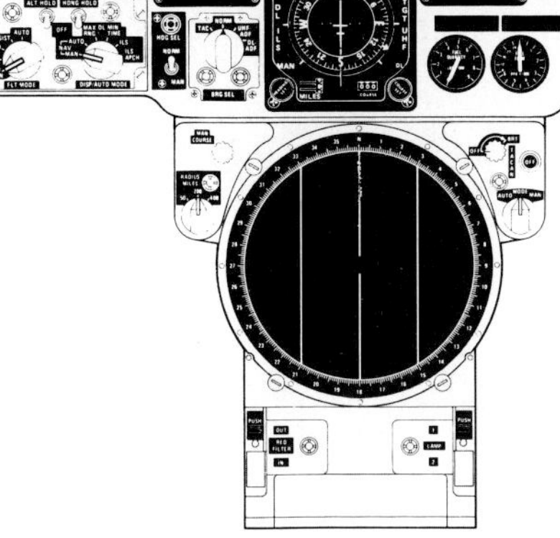

INSTRUMENT PANEL

1. Flight Modes Panel
2. Clock
3. UHF Remote Indicator
4. Tail Hook Down Button and Light
5. Computer Mode Indicator
6. Drag Chute Handle
7. Landing Gear Warning Light
8. Landing Gear Position Lights
9. Airspeed-Mach Indicator (AMI)
10. Attitude Director Indicator (ADI)
11. Maximum Maneuver Warning Light
12. External Tank Empty Lights
13. Radar Scope
14. Radar Scope Recorder
15. Standby Compass
16. Master Warning Light
17. Engine Fire Warning Light
18. Variable Ramp Warning Light
19. Engine Fire Warning Test Switch
20. Air Refueling Ready Light
21. Air Refueling Contact Light
22. Air Refueling Disconnect Light
23. Marker Beacon Light
24. Altitude-Vertical Velocity Indicator (AVVI)
25. Standby Attitude Indicator
26. Standby Altimeter
27. Standby Airspeed Indicator
28. Engine Pressure Ratio Gage
29. Exhaust Gas Temperature Gage
30. Tachometer
31. Fuel Quantity Gage Selector Switch
32. Nucleonic Oil Quantity Indicator
33. Radar/IR Selector Panel
34. Hydraulic Pressure-Low Warning Light
35. Canopy Unlocked Warning Light
36. Fuel Flow Indicator
37. Fuel Quantity Gage
38. Horizontal Situation Indicator (HSI)
39. Tactical Situation Display (TSD)
40. Bearing Selector Switch
41. Heading Selector Switch

LEFT CONSOLE

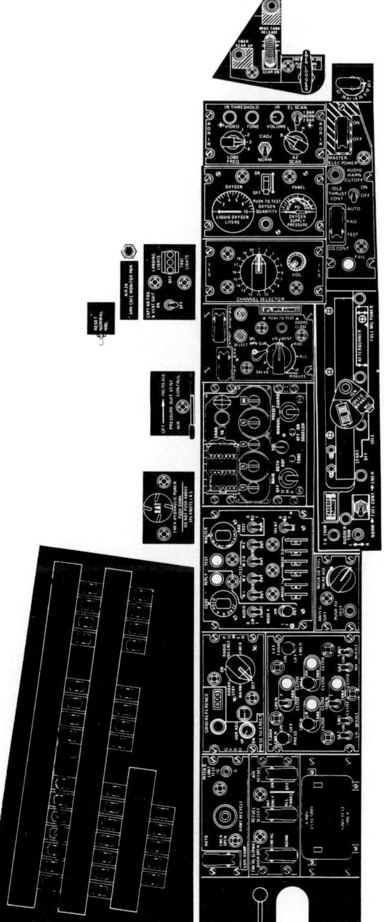

1. Landing Gear Emergency Up Button
2. Landing Gear Handle
3. External Wing Tanks Release Button
4. Landing Gear Emergency Extension Handle
5. Altitude Band Switch
6. Master Electrical Power Switch
7. Landing Gear Audio Warning Cutoff Button
8. Idle Thrust Control Switch
9. CG Control Switch
10. CG Transfer Test Failure Light
11. Radar/IR Control Panel
12. Oxygen Control Panel
13. ILS Channel Selector Panel
14. Armament Control Panel
15. Throttle Quadrant
16. Throttle
17. UHF Control Panel
18. Takeoff Trim Light
19. Fuel Control Switch
20. Rudder Trim Switch
21. IFF Control Panel
22. Mask Defog Rheostat
23. Anti-G Suit Test Button
24. MA-1 Power Control Panel
25. Fuel Control Panel
26. MA-1 Test Panel
27. Cabin Air Selector Handle
28. Air Refueling Panel
29. Variable Ramp Switch
30. Armament Recycle Button
31. Pitch G Limit Test Switch
32. Cockpit Left Fuse/Circuit Breaker Panel
33. Ram Air Turbine (RAT) Handle
34. Pressure Suit Control Handle
35. Reset/MBL Switch
36. Cockpit No-Fog and Ventilated Suit Switch
37. Landing and Taxi Light Switch
38. AIR-2A Arm/Safe/Monitor Power

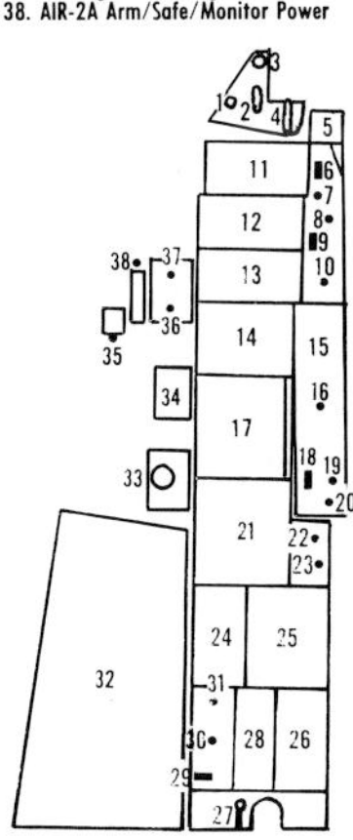

RIGHT CONSOLE

1. Master Warning Light Panel
2. Primary Hydraulic System Pressure Gage
3. Secondary Hydraulic System Pressure Gage
4. Oil Pressure Gage
5. IFF Caution Light
6. Generator Switch
7. No. 3 Fuel Tank Switch
8. Canopy Switch
9. Data Link Antenna Switch
10. Canopy Latch Handle
11. ATG Switch
12. Map Reading Light Switch
13. Warning Lights Test Button
14. Map Reading Light
15. Engine Anti-Ice Warning Test Button
16. Emergency AC Generator Switch
17. Windshield Anti-Icing, Antifog Switches
18. EGT Spread Button
19. Rain Removal Switch
19A. Altitude Warning Selector
20. Ejection Seat Ground Safety Pin Stowage
21. Cabin Temperature Control Knob
22. Pitot Heat Switch
23. Canopy Antifog Switch
24. Surface and Engine Anti-Icing Switch
25. Compass System Controller
26. Cockpit Right Fuse Circuit Breaker Panel
27. MATTS Switch
28. Map and Data Case
29. Lighting Control Powerstats
30. Lighting Control Panel
31. Data Link Control Panel
32. Auto-Navigation Homing Point Selector
33. TACAN Control Panel
34. Refrigeration Unit Switch
35. Cabin Air Selector Switch

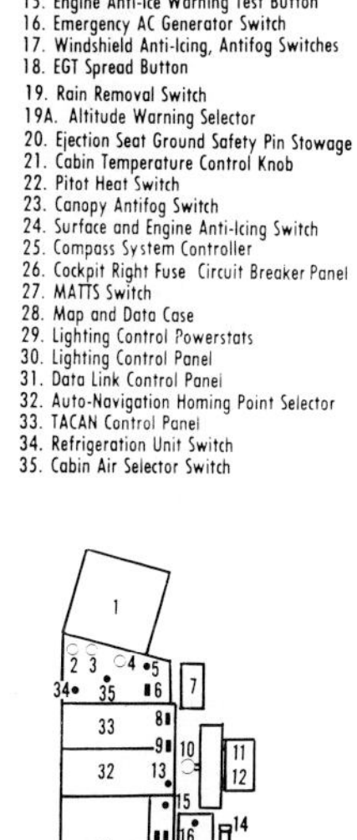

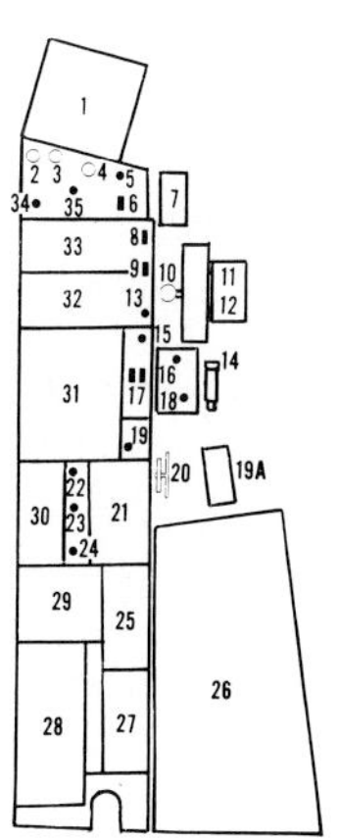

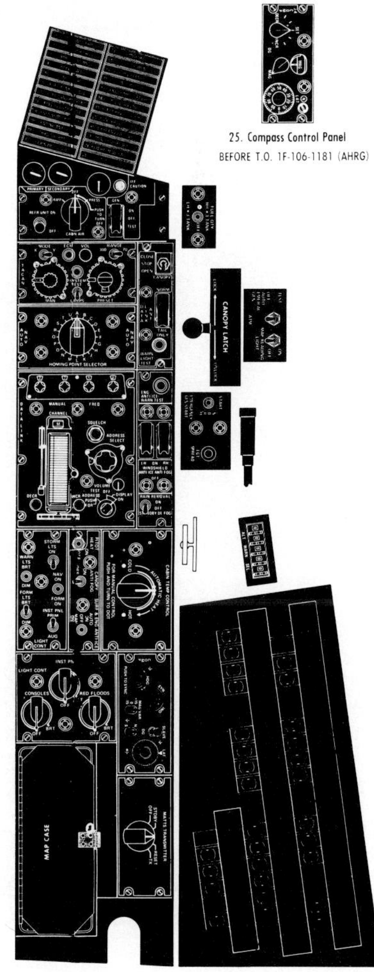

25. Compass Control Panel
BEFORE T.O. 1F-106-1181 (AHRG)

F-106A "SIX SHOOTER"

View showing the gun in place and the weapons bay doors closed.

This photo shows how the aft portion of the weapons bay doors were modified to fit around the gun fairing. (Barbier)

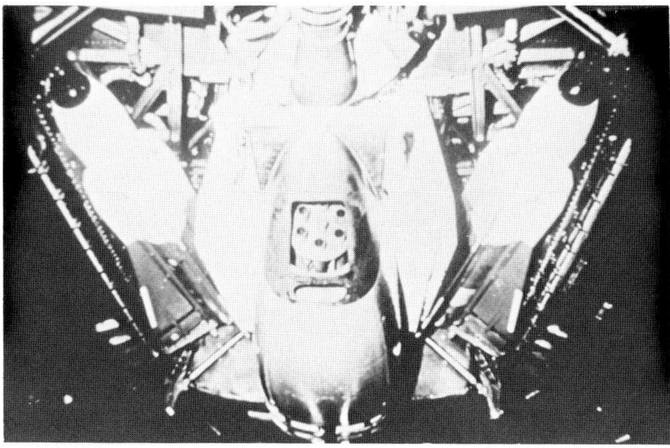

View looking up at the prototype gun installation located between the aft two Falcon missiles.(USAF)

Looking aft into the gun installation with the weapons bay doors open. (Barbier)

Two views of the gun installation removed from the aircraft.

Courtesy of the U.S.A.F.

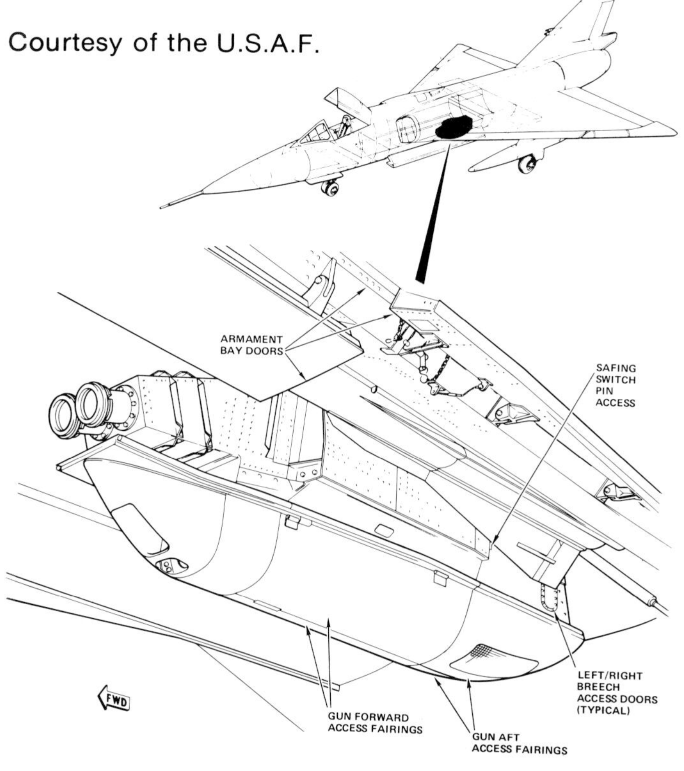

These two views, plus the one below, show the gun system removed from its structure. The ammunition drum, feed assembly, and barrels are clearly visible.

Drawing showing the details of the gun.

Courtesy of the U.S.A.F.

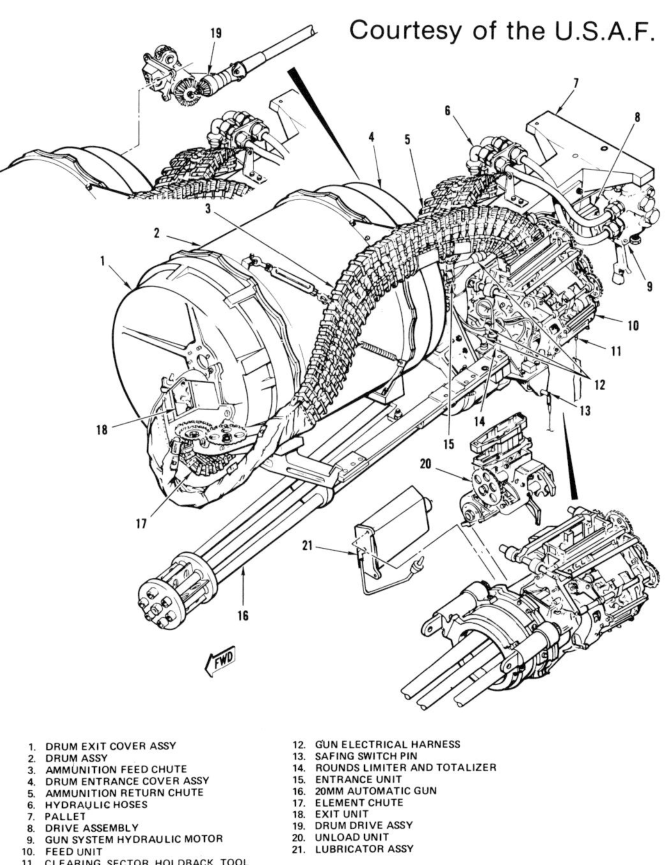

F-106A details continue on page 42.

F-106A, 90147, of the 318th FIS on a stopover at Offutt AFB in February 1980. *(Cockle)*

F-106A, 90081, from the 49th FIS, which is known as the Green Eagles.

F-106A, 80774, from the California Air National Guard. This photo, like the one to the left, was taken at William Tell, 1982.

Col. Dave Roeder (then a Captain) lifts the nose of 0-90135 as he rotates for take-off. The aircraft belongs to the 48th FIS. *(via Roeder)*

ACTIVE AIR FORCE COLORS

Above: The 2nd FIS was named the "Horney Horses" and operated their Delta Darts from Wurtsmith AFB, Michigan. (Strandberg via Flightleader)

Right: The 5th FIS flew the F-106 from Minot AFB, North Dakota, for more than a quarter of a century. (Flightleader)

Below: The 48th FIS was based at Langley AFB, Virginia, but the squadron also maintained a detachment at Homestead AFB, Florida, in the early 1970s. (Flightleader)

Above: The 49th FIS operated out of Griffiss AFB, New York. The markings on the tail denoted the squadron's "Green Eagles" nickname.

Left: Red, white, and blue markings were used by the "Black Panthers" of the 84th FIS. This unit first flew from Hamilton AFB, but later moved to Castle AFB. (Flightleader)

Below: Flying from K. I. Sawyer AFB, the "Red Bulls" of the 87th FIS painted a red bull motif on the tails of their F-106s. (Flightleader)

Above: The 95th FIS was based at Andrews AFB, Maryland, and was nicknamed "Mister Bones." (Munkasy)

Right: The 318th FIS flew the "Six" from McChord AFB, Washington. Although named the "Green Dragons," the unit used two shades of blue for its tail markings.

Below: The "Cave Tigers" of the 46th FIS were based at Grand Forks AFB, North Dakota. (Munkasy)

AIR NATIONAL GUARD COLORS

Left: The 194th FIS was assigned to the California ANG and was based at Fresno. In 1980 the unit used these colorful tail markings which included California's golden bear. (Flightleader)

Above: The 159th FIS was part of the Florida Air National Guard and based at Jacksonville International Airport. (Flightleader)

Below: Flying from Otis ANGB, the Massachusetts Air National Guard's 101st FIS began flying the "Six" in 1972.

Above: The 171st FIS was based at Selfridge AFB and was part of the Michigan Air National Guard. (Thurlow)

Right: The markings used by the Montana Air National Guard's 186th FIS never changed throughout the entire time the unit flew the F-106.

Below: Red markings adorned the Delta Darts flown by the 119th FIS. This unit was part of the New Jersey Air National Guard, and it was based at Atlantic City.

F-106 COCKPIT COLORS

This is the front instrument panel in an F-106B with the integrated flight instrument system. The cockpit in an F-106A would also look like this. Compare these instruments to the "round eye" style instruments illustrated on pages 56 and 57. (Mason)

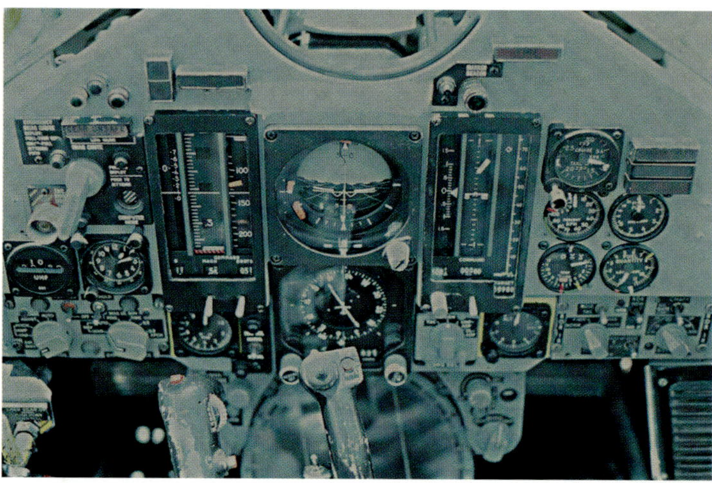

Details of the rear panel in an F-106B with the integrated flight instrument system are revealed here. The vertical "tape" style instrument to the left indicated airspeed, and the one to the right displayed altitude, and vertical speed. (Mason)

Details on the left console in the front cockpit are visible in this photograph. The throttle was mounted on a quadrant next to the console rather than being on the console itself. (Mason)

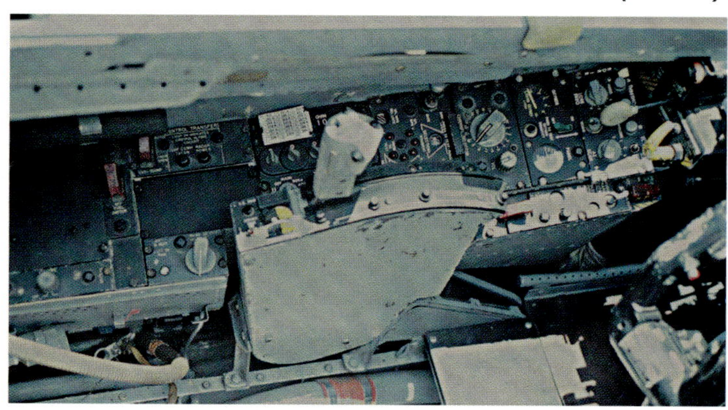

The throttle quadrant remained the same in the rear cockpit, but the details on the console were fewer in number. The items in the cockpits are identified on pages 60 and 61. (Mason)

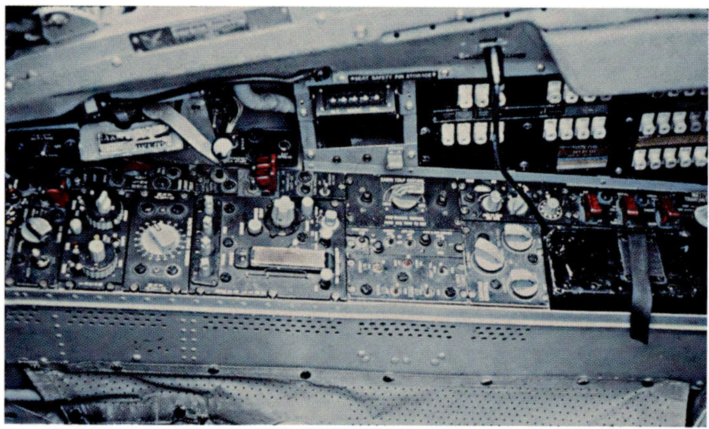

The master warning light panel was at the forward end of the right console in the front cockpit. Other items on this side of the cockpit included the gages for the oil and hydraulic systems, the canopy latch handle, and the map reading light. (Mason)

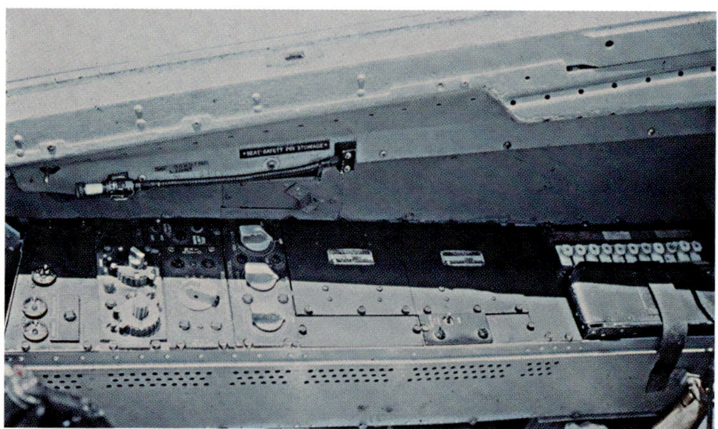

The major items on the right console in the front cockpit were repeated in the rear cockpit, but most of the non-essential features were deleted. The six photographs on this page were taken with the ejection seats removed to better reveal the details. (Mason)

RADAR & AVIONICS COLORS

The radome has been removed and the access doors covering the avionics bays have been opened for maintenance on this in this front view of an F-106A. (Mason)

A right front view shows the radar mount and some of the avionics equipment to good effect. (Mason)

This left front view provides a good look at the radar antenna dish and its feed horn. (Mason)

Items located inside the right avionics bay are revealed in this view. The interior of the bay is painted with green primer. (Mason)

This photograph of the left avionics bay shows many of the black boxes associated with the Hughes radar equipment. (Mason)

MISSILE ARMAMENT COLORS

Above: The standard load of four AIM-4 Falcon missiles are loaded on this F-106A. The front two missiles are AIM-4F radar-guided Falcons, and the rear two missiles are AIM-4G infrared-guided Falcons. (Mason)

Left: This rear view provides a good look at the extended launch rails. The white areas on the leading edges of the fins are the fuses, and these must contact the target to detonate the missile's warhead. (Mason)

When the use of nuclear weapons was authorized by the President of the United States, an unguided AIR-2 Genie rocket with a tactical nuclear warhead could be loaded between the rear two Falcon missiles. The blue color on the body of this Genie indicates that it is an inert training version of the weapon. (Mason)

This underside view looks forward into the weapons bay and shows the Genie rocket loaded between the aft two Falcon missiles. Their launch rails are in the retracted position. Late in the operational life of the F-106A, the Vulcan gun pod could be installed in place of the Genie as illustrated on pages 30 and 31. (Mason)

ARMAMENT

An F-106A from the 5th FIS is shown launching an AIR-2A Genie rocket. ***(USAF)***

The missile armament of the F-106 included four AIM-4 Falcons and one AIR-2A Genie unguided nuclear rocket. Four AIM-4G or four AIM-4F Falcons could be carried, but usual armament consisted of two of each type. The -F Falcon is radar guided, and the -G is the infrared guided version. When both types of missiles are carried, the -Fs are on the forward rails and the -Gs are on the aft rails. When salvoed in pairs, and all four missiles are launched at a single target, the rear missiles are launched first, and are then followed by the forward missiles. The use of two of each type of missile increases the PK (probability of kill), and the use of two types of guidance systems makes the missile attack harder to defend against. But when we use the Genie - well, that's like killing a with a sledge hammer."

F-106 Weapons Data Tables

M61A1 20mm Vulcan Cannon, General Electric

Firing rate fixed at 4,500 rounds/min.
Ammunition capacity = 650 rounds
Total System Weight, including gun, pallet, feed system and ammunition = 1,059 lb.

AIM-4F Super Falcon Guided Missile, Hughes Aircraft Corp.

Semi-Active Radar Guidance	Dimensions
High explosive warhead	
Launch Weight = 150 lb.	Length = 7′ 2″
Max Speed = Mach 3	Width = 2′
Max Range = 7 miles	(Body) = 6.6″

AIM-4G Super Falcon Guided Missile, Hughes Aircraft Corp.

Infrared Guidance	Dimensions
High Explosive Warhead	
Launch Weight = 145 lb.	Length = 6′ 9″
Max Speed = Mach 3	Width = 2′
Max Range = 7 Miles	(Body) = 6.6″

AIR-2A Genie Unguided Rocket, McDonnell/Douglas

Unguided	
Nominal 1.5 kiloton nuclear warhead	
Launch Weight = 837.7	Length = 9′
Max Speed = Mach 3	Width = 3.3′
Max Range = 6 miles	(Body) = 1′ 6″

F-106A WEAPONS BAY DETAIL

Weapons bay looking aft. Note the angle of the doors, one to the other. The missiles are of the training variety. ***(Spering)***

F-106A weapons bay looking forward. ***(Spering)***

In the F-106A, the forward two missile rails are joined by a crossbridge support assembly. This photo shows this assembly and the launch rails in a view looking forward. (Spering)

Looking aft at the forward end of the crossbridge support assembly.

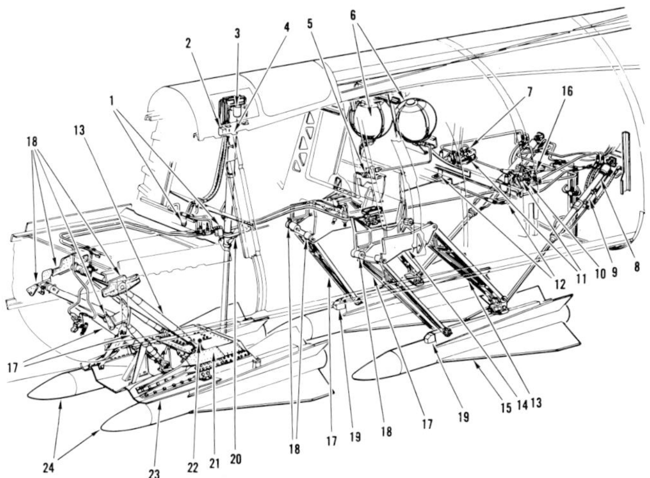

1. FORWARD LAUNCHER RETRACT SNUBBERS
2. FORWARD LAUNCHER SOLENOID SELECTOR VALVE
3. FORWARD LAUNCHER UPLOCK LIMIT SWITCH
4. FORWARD LAUNCHER ACTUATOR
5. 1500 PSI PRESSURE REGULATOR
6. 3000 PSI PNEUMATIC SYSTEM FLASKS
7. AFT LAUNCHER UPLATCH LIMIT SWITCH
8. AFT LAUNCHER ACTUATORS
9. PRESSURE OPERATED SELECTOR VALVES (INTEGRAL PART OF ACTUATOR)
10. AFT UPLATCH ACTUATORS
11. AFT LAUNCHER UPLATCH ASSEMBLIES
12. AFT LAUNCHER RETRACT SNUBBERS
13. AFT LAUNCHER UPPER DRAG ARMS
14. LAUNCHER TRUNNION BOXES
15. AFT MISSILES
16. AFT LAUNCHER SOLENOID SELECTOR VALVE
17. LAUNCHER LOWER DRAG ARMS
18. DRAG ARM TRUNNIONS
19. AFT LAUNCHER ASSEMBLIES
20. BLAST DEFLECTOR
21. FORWARD ARMAMENT CROSSBRIDGE SUPPORT ASSEMBLY
22. SNUBBER BUMPER BOLTS (TWO)
23. FORWARD LAUNCHERS AND SUPPORT ASSEMBLY
24. FORWARD MISSILES

Courtesy of the U.S.A.F.

View looking forward and up showing the supports and weapons bay area above the crossbridge support assembly.

View looking aft and up into the rear portion of the missile bay.

Aft left missile launch rail in the extended position.

PYLONS

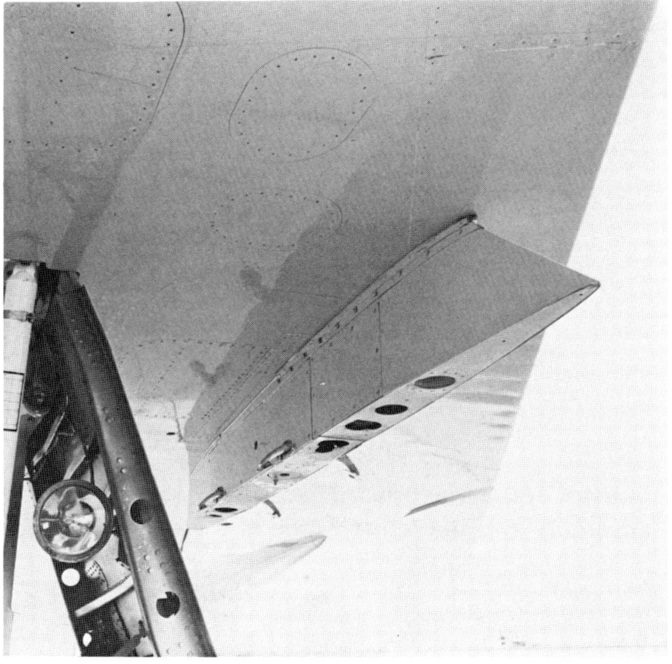

***Above left:* Standard 360 gallon external fuel tank attached to the left wing pylon.**

***Above right:* View looking at the left wing pylon from below and in front. Note the holes and the anti-sway braces.**

***Left:* Close-up of the left pylon with the fuel tank attached. Note how the anti-sway braces fit on the tank.**

***Below left and right:* Two views of the right wing pylon without the tank in place.**

Right wing pylon with a special adapter and an Air Combat Maneuvering Instrumentation (ACMI) pod attached. This is an example of one of the rare instances when the pylons are not used to carry fuel tanks. (Spering)

View showing the rear of the tank looking forward toward the pylon. Note that the tank is not pointed at the rear, but is blunt. Many tanks have a data plate located on the end as shown, but it is not always present.

Right wing pylon shown close up with the tank in place. Note the slot in the wing.

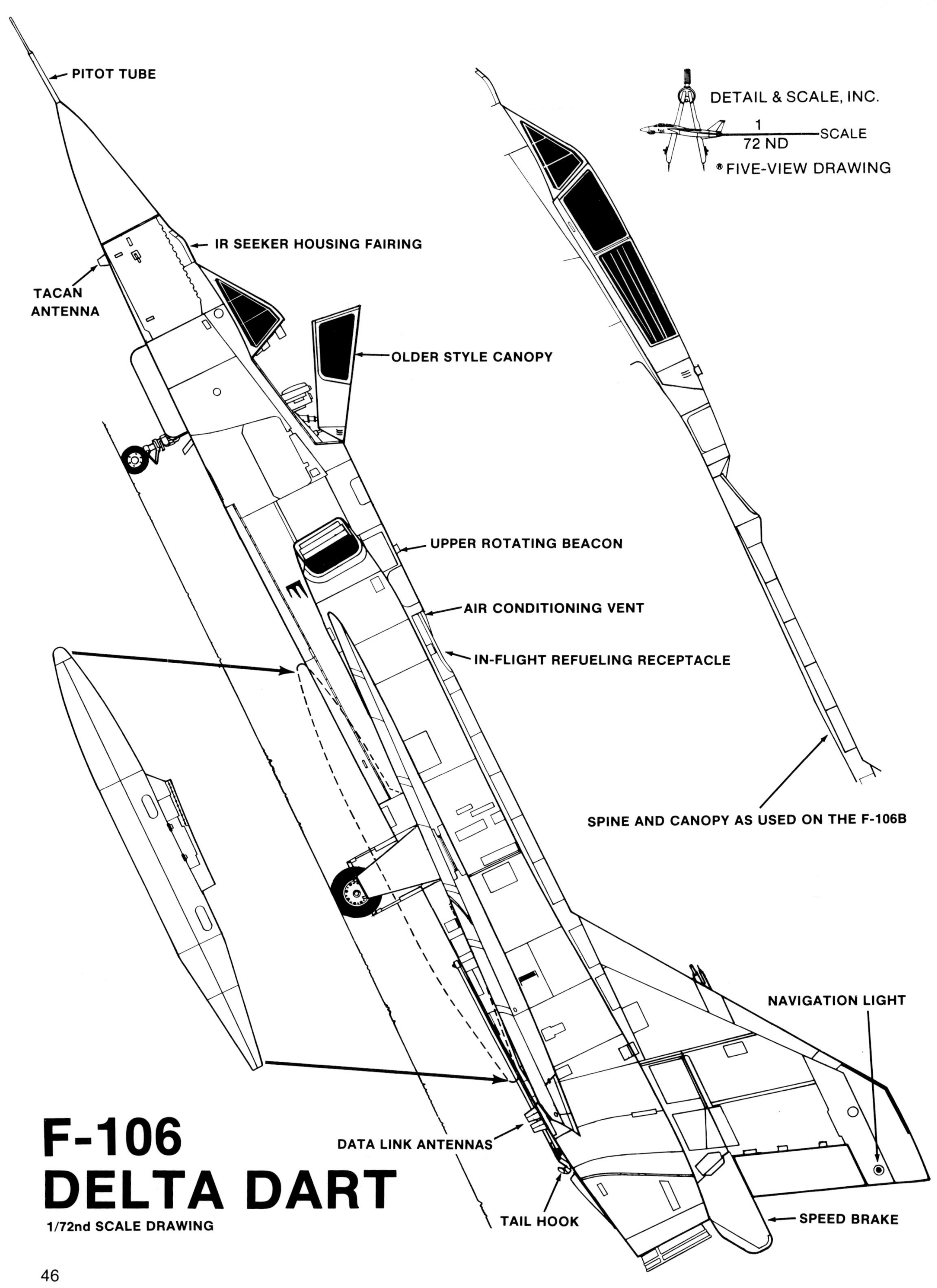

F-106 DELTA DART

1/72nd SCALE DRAWING

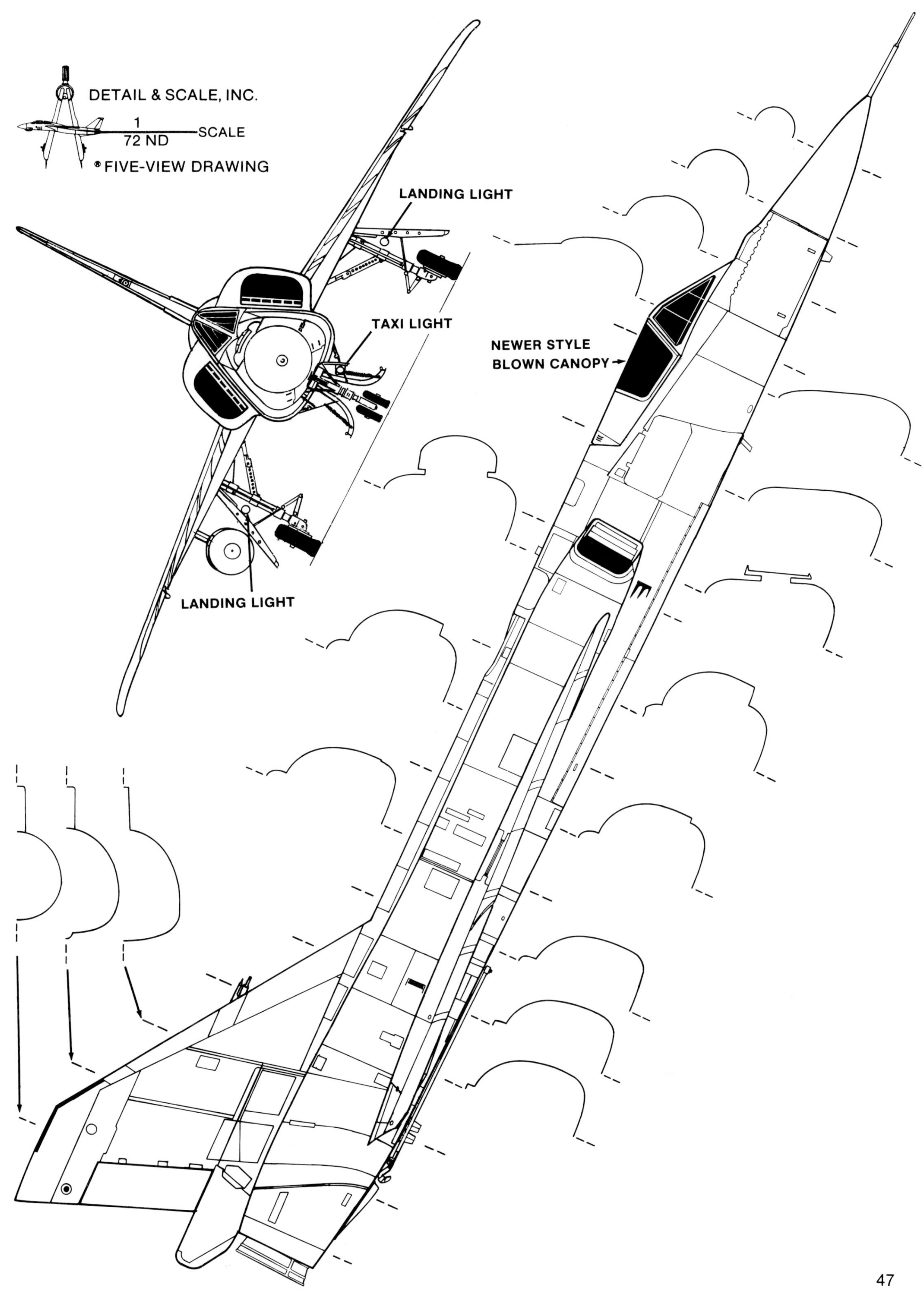
DETAIL & SCALE, INC.
1
72 ND
SCALE
® FIVE-VIEW DRAWING
LANDING LIGHT
TAXI LIGHT
NEWER STYLE
BLOWN CANOPY
LANDING LIGHT

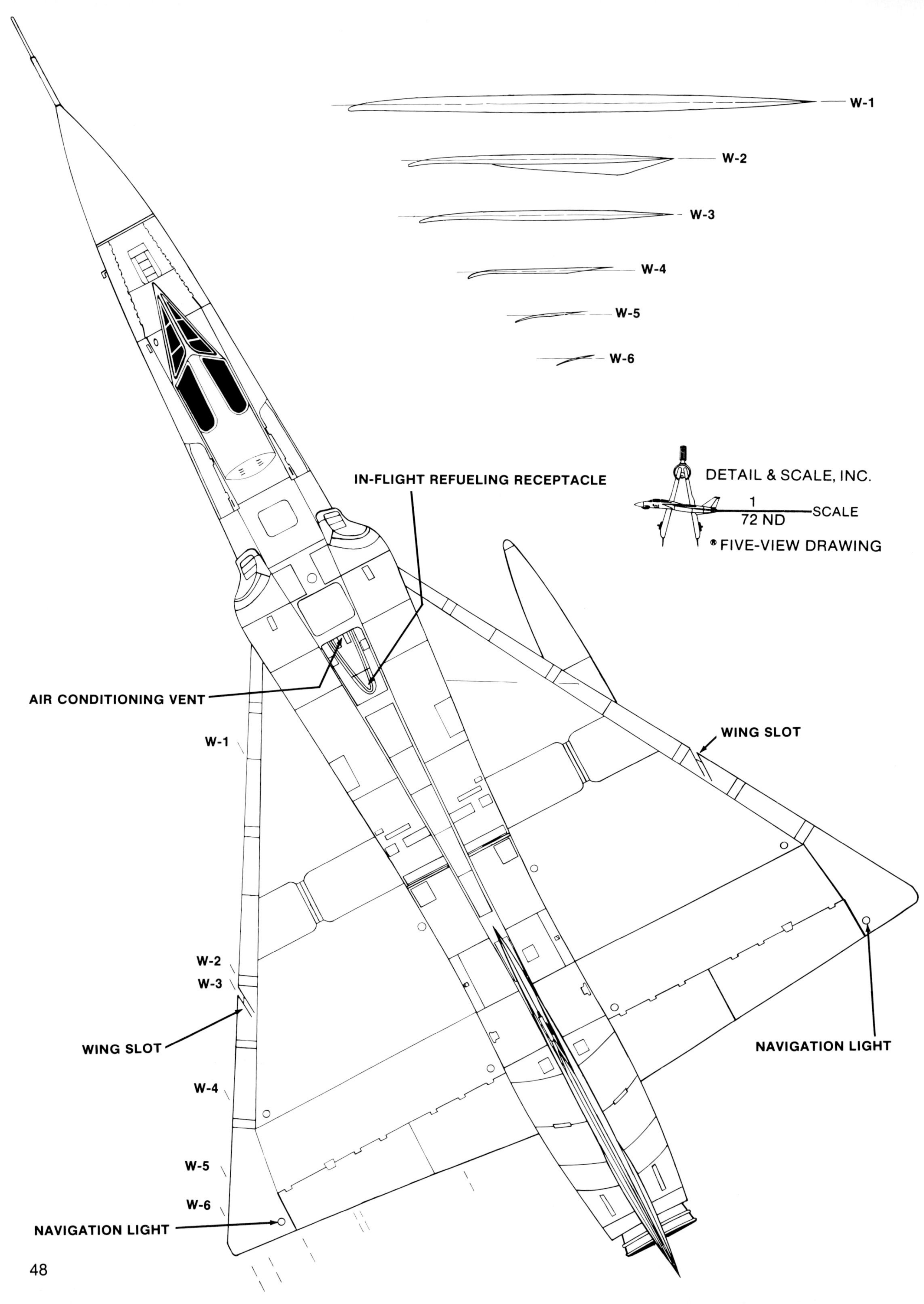
W-1
W-2
W-3
W-4
W-5
W-6
IN-FLIGHT REFUELING RECEPTACLE
DETAIL & SCALE, INC.
1
72 ND
SCALE
® FIVE-VIEW DRAWING
AIR CONDITIONING VENT
W-1
WING SLOT
W-2
W-3
WING SLOT
NAVIGATION LIGHT
W-4
W-5
W-6
NAVIGATION LIGHT

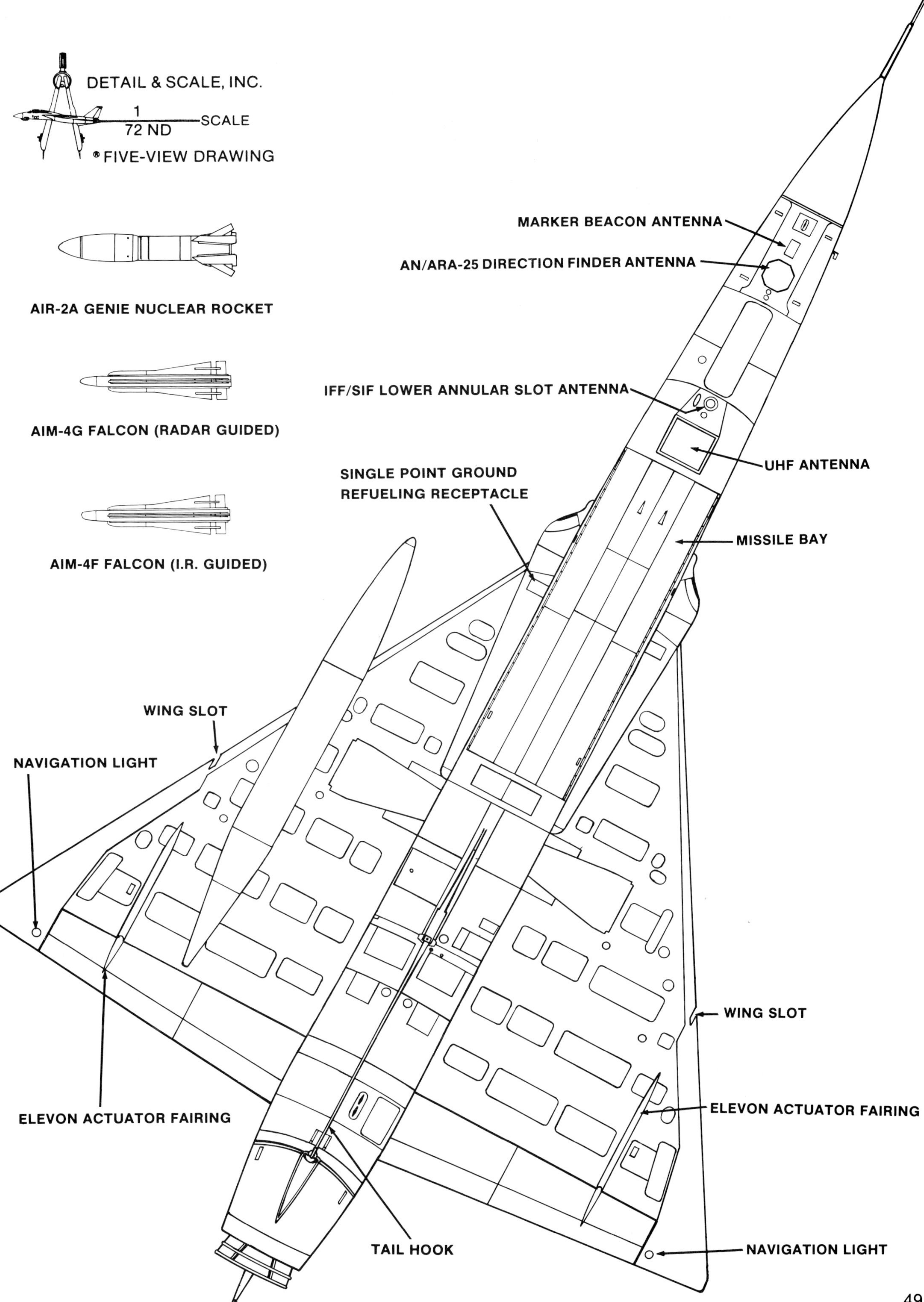
DETAIL & SCALE, INC.
1
72 ND
SCALE
® FIVE-VIEW DRAWING
AIR-2A GENIE NUCLEAR ROCKET
AIM-4G FALCON (RADAR GUIDED)
AIM-4F FALCON (I.R. GUIDED)
MARKER BEACON ANTENNA
AN/ARA-25 DIRECTION FINDER ANTENNA
IFF/SIF LOWER ANNULAR SLOT ANTENNA
UHF ANTENNA
SINGLE POINT GROUND REFUELING RECEPTACLE
MISSILE BAY
WING SLOT
NAVIGATION LIGHT
WING SLOT
ELEVON ACTUATOR FAIRING
ELEVON ACTUATOR FAIRING
TAIL HOOK
NAVIGATION LIGHT

DIMENSION DATA

DIMENSION	ACTUAL	1/72nd SCALE	1/48th SCALE	1/32nd SCALE
Length	70′ 8.78″	11.79″	17.68″	26.52″
Wingspan	38′ 3.5″	6.38″	9.57″	14.35″
Height	20′ 3.3″	3.38″	5.07″	7.60″
Wheel Track	24′ 1.5″	4.02″	6.03″	9.05″
Wheel Tread	15′ 5.64″	2.58″	3.87″	5.80″

Note: Dimensions for the F-106B are the same as those for the F-106A.

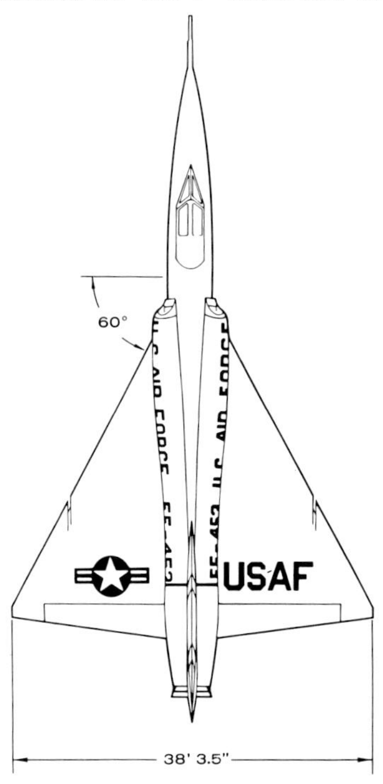

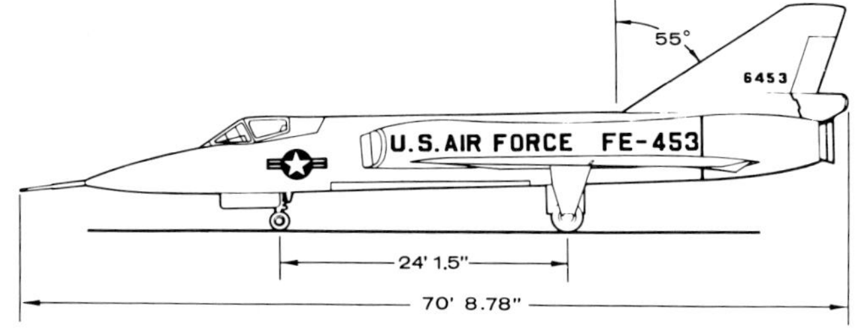

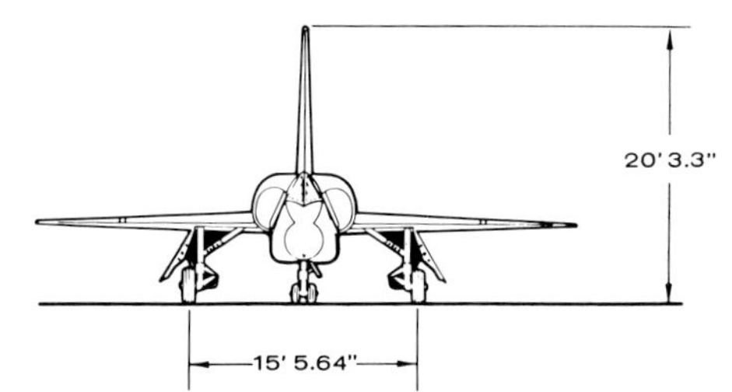

WING AREA .... 697.83 sq. ft.

ASPECT RATIO 2.2

MEAN AERODYNAMIC CHORD .... 285.1 in.

WING SECTION .... NACA 0004-65 (mod)

PRODUCTION BLOCKS & SERIALS

F-106A

BLOCK	NO. BUILT	SERIAL NUMBERS
F-106A-1-CO	17	56-451 - 56-467
F-106A-1-CO	18	57-229 - 57-246
F-106A-65-CO	3	57-2453 - 57-2455
F-106A-70-CO	5	57-2456 - 57-24760
F-106A-75-CO	5	57-2461 - 57-2465
F-106A-80-CO	12	57-2466 - 57-2477
F-106A-85-CO	8	57-2478 - 57-2485
F-106A-90-CO	21	57-2486 - 57-2506
F-106A-95-CO	13	58-759 - 58-771
F-106A-100-CO	27	58-772 - 58-798
F-106A-105-CO	30	59-001 - 59-030
F-106A-110-CO	29	59-031 - 59-059
F-106A-120-CO	27	59-060 - 59-086
F-106A-125-CO	25	59-087 - 59-111
F-106A-130-CO	24	59-112 - 59-135
F-106A-135-CO	13	59-136 - 59-148

F-106B

BLOCK	NO. BUILT	SERIAL NUMBERS
F-106B-31-CO	11	57-2507 - 57-2517
F-106B-35-CO	2	57-2518 - 57-2519
F-106B-40-CO	3	57-2520 - 57-2522
F-106B-45-CO	4	57-2523 - 57-2526
F-106B-50-CO	5	57-2727 - 57-2531
F-106B-55-CO	10	57-2532 - 57-2541
F-106B-60-CO	6	57-2542 - 57-2547
F-106B-65-CO	5	58-900 - 58-904
F-106B-75-CO	11	59-149 - 59-159
F-106B-80-CO	6	59-160 - 59-165

TOTALS: F-106A, 277
F-106B, 63

Notes: All aircraft in the F-106A-1-CO blocks were improved to later standards and redesignated F-106A-31-CO. F-106Bs 57-2508 - 57-2515 were improved to later standards and redesignated F-106B-81-CO.

THE F-106B

This F-106B from the Montana Air National Guard appears to be in mint condition. As a rule, all F-106s usually had a clean, well maintained, glossy finish. ***(Cockle)***

The F-106B was originally designated the TF-106A, but because the aircraft retained the full combat capability of the F-106A, it was redesignated F-106B. With the addition of the rear cockpit, the aft avionics had to be moved to the forward center portion of the weapons bay, and this changed the configuration of the bay.

The length of the airframe was not changed with the addition of the rear cockpit. The spine was heightened at the front end to taper past the refueling receptacle to the tail. A large single-piece canopy covered both cockpits.

The first F-106B made its maiden flight on April 9, 1958, and was quickly accepted by the Air Force. Initial operational capability was in July 1960. In all, a total of sixty-three F-106Bs were accepted by the Air Force with the last delivered in late 1960. Flyaway cost was 4.9 million dollars per aircraft.

As noted earlier, the F-106B received all of the updates and modifications as the F-106A, with the exceptions of the clear canopy and the internal gun.

F-106Bs have been used for several test programs wearing NASA colors. These include a very unusual three-engine aircraft, and a lightning strike aircraft as shown in the color section. Details of the F-106B are covered on the next few pages, and emphasis is given to the differences between it and the F-106A. Additionally, complete cockpit coverage is shown of an F-106B with the "round eye" instrument panels. The F-106A and F-106B originally had this style of instruments before the tape styles were fitted in later aircraft. It should be noted that the earlier F-106s were not retrofitted with the tape instruments, and that the gun installation was not added to any F-106As with the older "round eye" instruments. The instrument panel in a "round eye" F-106A would be the same as that shown for the front cockpit of the F-106B.

F-106B, 72524, shown in the markings of the Massachusetts Air National Guard. ***(via Barbier)***

Head-on view taken from slightly above an F-106B from the California Air National Guard. ***(Cockle)***

The panel hinged down is the UHF Antenna on the F-106B. The F-106A has the same arrangement.

View looking up into the area above the open UHF antenna. The round antenna to the left is the IFF Lower Annular Slot Antenna.

F-106B, 58-902 had RHAW gear installed and tested for about one and one-half years in the early 1970s. The photo on the right shows a close-up of the antenna installation on the vertical tail which remains there today.

F-106B TECHNICAL DATA

Mission and Description

Navy Equivalent: None — Mfr's Model: 8-27

The principal mission of the F-106B is to function as a pilot proficiency training while maintaining full tactical capabilities for the interception and destruction of hostile aircraft and missiles. The F-106B has all-weather and day or night characteristics.

This airplane incorporates a delta wing with a cambered leading edge extending from wing root to wing tip and swept tail surface. Control surfaces are power operated.

The fuel system is pressurized, air is bled from the engine compressor section and is used to pressurize the fuel tanks to reduce fuel evaporation.

The airplane has the capability for air refueling from flying boom equipped tankers.

The armament is located in a bay in the bottom of the fuselage. The AIM missiles are extended below this section for firing and the AIR-2A rocket is ejected from the bay by an explosive change. Firing of the armament is either manual or automatic. The components of the AN/ASQ-25 (MA-1 equivalent) Aircraft and Weapons Control System provides automatic radar searching and tracking, directs the airplane on a lead-collision attack and automatically fires the armament.

External fuel tanks can be added to increase range. The tanks can be refueled in flight and need not be jettisoned for combat since they do not restrict airplane speed or load factor when empty.

Development

Basically the same as the F-106A except for redesign of the cockpit to accommodate two pilots, redesign of the fuselage fuel tank area and installation of a comparable electronic system.

First flight (prototype)	Apr 58
First flight (production)	Oct 58
Production status	Production Completed

POWER PLANT

Nr & Model	(1) J75-P-17
Mfr	Pratt & Whitney
Engine Spec Nr	A-2625
Type	Axial
Length	237.6"
Diameter	44.25"
Weight (dry)	5875 lb
Tail Pipe	Auto, Two-Position
Augmentation	Afterburning

ENGINE RATINGS

S. L. Static	LB	† RPM	MIN
Max:	*24,500	6440/8940	5
Mil:	16,100	6440/8940	15
Nor:	13,700	6080/8700	Cont

* With afterburner operating

† First figure represents RPM of low pressure spool while the second that of the high pressure spool.

FUEL

Location	Nr Tanks	Gal
Fuselage	1	176
Wg, internal	8	1267
Wg, ext, drop	2	716
Transfer lines		7
	Total	2166
Grade		JP-4
Specification		MIL-T-5624

OIL

Engine	(tot) 4.5
Specification	MIL-L-7808

ELECTRONICS

Interceptor System, Aircraft and Weapons Control, AN/ASQ-25

(Hughes Aircraft Corp)

WEIGHTS

Loading	Lb	L. F.
Empty	25,696(A)	
Basic	26,200(A)	
Design	34,937	6.0
Combat	*34,482	6.0
Max T.O.	†42,720	3.9
Max Land	‡36,834	2.0

(A) Actual
* For basic mission
† Limited by space
‡ Limited by design

ROCKETS

Nr	Type	Location
1	AIR-2A plus	Fuselage
4	AIM-4F or	Fuselage
4	AIM-4G or	Fuselage
2	AIM-4F	Fuselage
2	AIM-4G	Fuselage

Courtesy of the U.S.A.F.

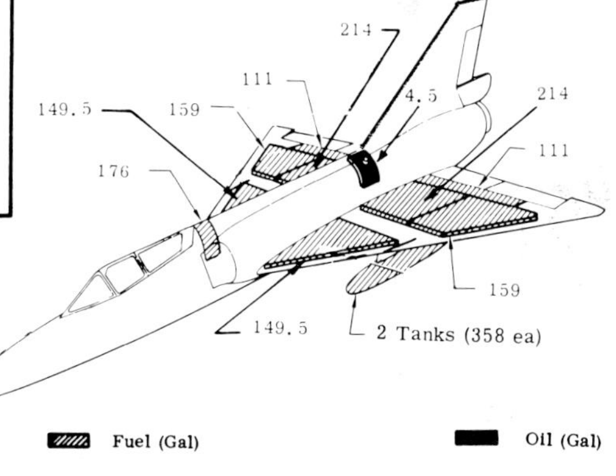

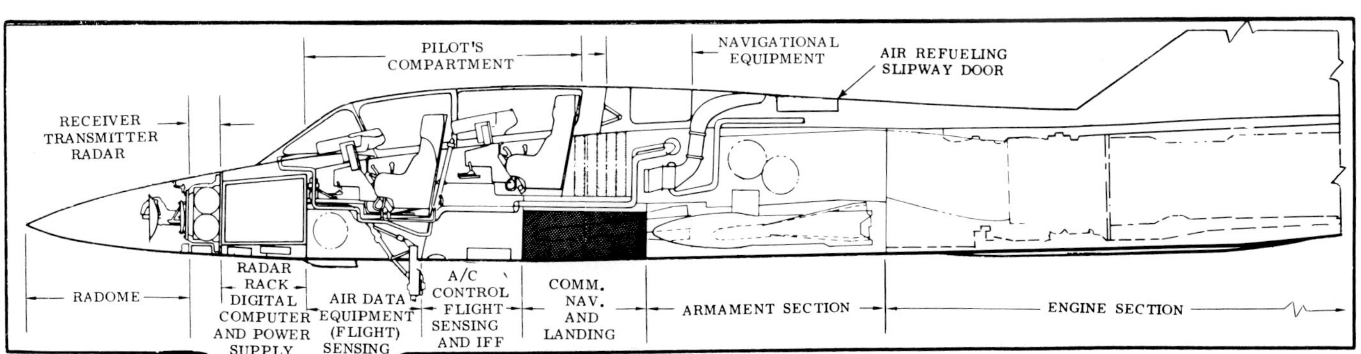

F-106B PERFORMANCE DATA

Loading and Performance—Typical Mission

CONDITIONS			MAXIMUM INTERNAL FUEL MISSION			EXTERNAL FUEL MISSION	
			POINT INTERCEPT I	AREA INTERCEPT II	FERRY RANGE III	AREA INTERCEPT IV	FERRY RANGE V
TAKEOFF WEIGHT		(lb)	37,552	37,552	36,129	42,720	41,297
Fuel at 6.5 lb/gal (grade JP-4)		(lb)	9425	9425	9425	14,079	14,079
Military load (missiles)	(5)	(lb)	594	594	—	594	—
Military load (rockets)	(6)	(lb)	829	829	—	829	—
Wing loading		(lb/sq ft)	54	54	52	61.5	59.4
Minimum speed (power off)	(9)	(kn)	150	150	150	150	150
Takeoff ground run	(1)	(ft)	3650	3650	3350	4650	4350
Takeoff to clear 50 ft	(1)	(ft)	5400	5400	5100	6600	6250
Rate of climb at SL		(ft/min)	38,000 (1) (8)	8600 (2) (8)	8700 (2)	6800 (2) (8)	6875 (2)
Time: SL to 40,000 ft	(8)	(min)	3.4 (1) (7)	8.2 (2)	7.6 (2)	15.5 (2)	13.6 (2)
Time: SL to 50,000 ft	(8)	(min)	6.3 (1) (7)	16.4 (2) (11)	15.8 (2) (11)	19.0 (2) (11)	18.7 (2) (11)
Service ceiling (100 ft/min)		(ft)	51,200 (1) (8)	44,100 (2) (8)	43,200 (2)	41,200 (2) (8)	39,800 (2)
COMBAT RANGE		(n. mi.)	—	—	977	—	1600
COMBAT RADIUS		(n. mi.)	—	377	—	563	—
Average cruise speed		(kn)	—	516	516	516	516
Initial cruising altitude		(ft)	—	38,200	39,000	36,000	36,800
Final cruising altitude		(ft)	—	41,000	41,700	40,600	41,300
Total mission time		(hr)	—	1.54	1.82	2.26	3.38
TOTAL MISSION TIME	(7)	(hr)	1.79	—	—	—	—
Intercept altitude		(ft)	50,600	—	—	—	—
COMBAT WEIGHT		(lb)	34,482	32,822	28,286	35,297	29,033
Combat altitude		(ft)	50,600	51,450	41,700	50,000	41,300
Combat speed	(1)	(kn)	588	588	—	588	—
Combat climb	(1)	(ft/min)	500	500	9500	500	9500
Combat ceiling (500 ft/min)	(1)	(ft)	50,600	51,450	53,200	50,000	53,000
Service ceiling (100 ft/min)	(2)	(ft)	51,100 (1)	45,300	46,700	43,800	46,400
Maximum rate of climb at SL	(1)	(ft)	40,000	42,000	48,500	39,100	47,500
Maximum speed at 35,000 ft	(1) (10)	(kn)	1153	1153	1153	1153	1153
Basic speed at 50,000 ft	(1)	(kn)	1061	1089	1100	1055	1054
LANDING WEIGHT		(lb)	29,377	29,709	28,286	30,456	29,033
Ground roll at SL		(ft)	4940	5000	4770	5100	4890
Ground roll (auxiliary brake)	(12)	(ft)	3620	3670	3440	3770	3550
Total from 50 ft		(ft)	6370	6420	6200	6520	6310
Total from 50 ft (auxiliary brake)	(12)	(ft)	5050	5100	4870	5200	5000

NOTES

(1) Maximum thrust
(2) Military thrust
(3) Deleted
(4) With 716 gallons external fuel
(5) Four AIM-4F or 4G missiles
(6) One AIR-2A rocket
(7) Includes time for takeoff and acceleration to climb speed
(8) Considers weight reduction due to fuel used.
(9) Onset of heavy buffet
(10) Design speed limit (M = 2.0)
(11) Time to service ceiling
(12) 14.5 ft (flat diameter) drag chute plus speed brakes.

PERFORMANCE
Data source: Flight Test.

Courtesy of the U.S.A.F.

F-106B CANOPY DETAIL

***Above left:** F-106B canopy viewed directly from the front.* *(Spering)*

***Above right:** F-106B canopy details as seen from the front right. Note the raising and lowering cylinder and the details of the canopy rails.* *(Barbier)*

***Right:** Canopy viewed from the left side. Note the black radar scope located between the rails.* *(King)*

***Below:** Right side view of the open canopy on an F-106B.* *(Spering)*

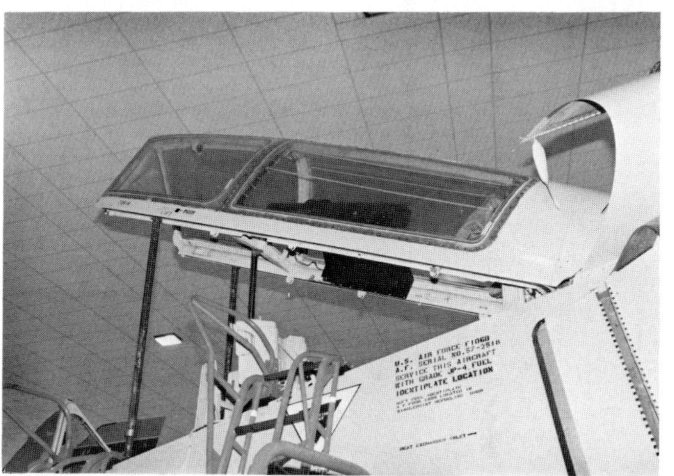

F-106B "ROUND EYE" COCKPIT DETAILS

The original "round eye" instrument panel in an F-106B.

Left console in the front cockpit of a "round eye" F-106B.

Right console detail in the front cockpit.

Above left: "Round eye" instrument panel in the rear cockpit of an F-106B. F-106s with the "round eye" instruments were not retrofitted with the tape type instruments.

Above right: Looking up at the radar scope for the rear cockpit which is mounted between the rails of the canopy. The canopy was open for this photo.

Below left: Left console and throttle in the rear cockpit.

Below right: Right console in the rear cockpit. The seat has been removed from the aircraft for servicing. For detail drawings of these cockpits see the next two pages.

F-106B COCKPIT LAYOUTS

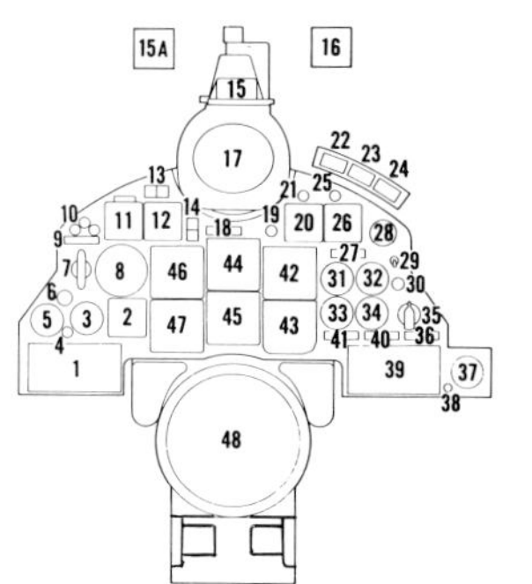

F-106B cockpit -typical-forward

CONVENTIONAL INSTRUMENT DISPLAY

Courtesy of the U.S.A.F.

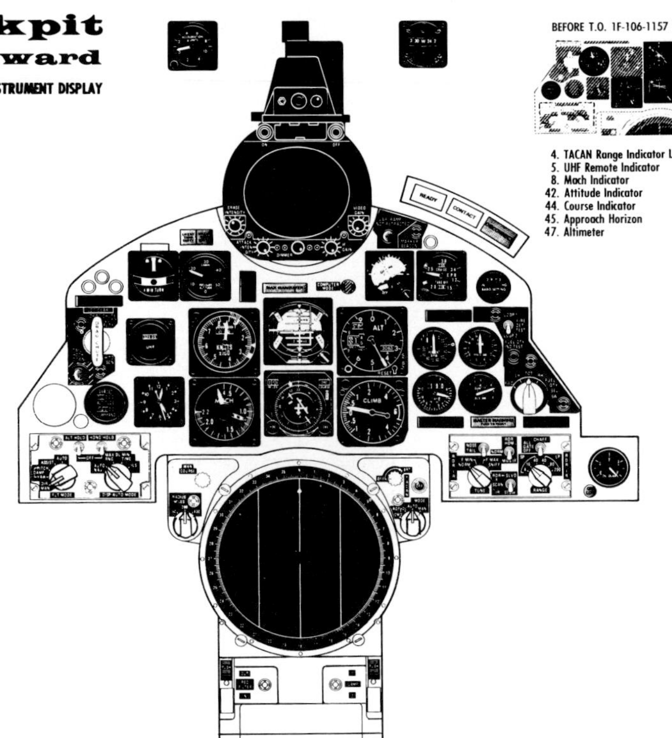

BEFORE T.O. 1F-106-1157 (DME)

4. TACAN Range Indicator Light
5. UHF Remote Indicator
8. Mach Indicator
42. Attitude Indicator
44. Course Indicator
45. Approach Horizon
47. Altimeter

INSTRUMENT PANEL

1. Flight Modes Panel
2. Clock
3. Command and Target Altitude Indicator
4. Blank
5. Blank
6. Tail Hook Down Button and Light
7. Drag Chute Handle
8. UHF Remote Indicator
9. Landing Gear Warning Light
10. Landing Gear Position Lights
11. Turn-and-Slip Indicator
12. Cabin Pressure Altitude Gage
13. External Tank Empty Lights
14. Bailout Warning Lights
15. Radar Scope Recorder
15A. Accelerometer
16. Standby Compass
17. Radar Scope
18. Maximum Maneuver Warning Light
19. Computer Mode Indicator
20. Standby Attitude Indicator
21. Variable Ramp Warning Light
22. Air Refueling Ready Light
23. Air Refueling Contact Light
24. Air Refueling Disconnect Light
25. Marker Beacon Light
26. Engine Pressure Ratio Gage
27. Engine Fire Warning Light
28. Barometer Setting Control
29. Engine Fire Warning Test Switch
30. Fuel Quantity Gage Test Button
31. Tachometer
32. Fuel Flow Indicator
33. Exhaust Gas Temperature Gage
34. Fuel Quantity Gage
35. Fuel Quantity Gage Selector Switch
36. Hydraulic Pressure-Low Warning Light
37. Nucleonic Oil Quantity Indicator
38. Oil Quantity-Low Light
39. Radar/IR Selector Panel
40. Master Warning Light
41. Canopy Unlocked Warning Light
42. Altimeter
43. Vertical Velocity Indicator
44. Attitude Director Indicator
45. Horizontal Situation Indicator
46. Airspeed-Angle of Attack Indicator
47. Mach Indicator
48. Tactical Situation Display (TSD)

LEFT CONSOLE

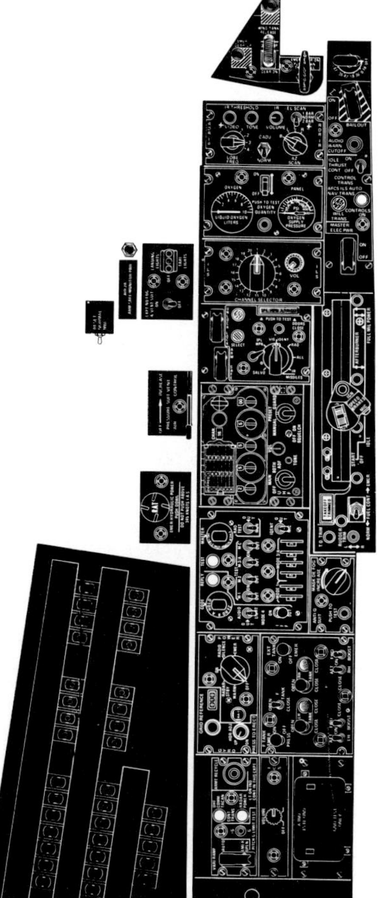

1. Landing Gear Emergency Up Button
2. Landing Gear Handle
3. External Wing Tanks Release Button
4. Landing Gear Emergency Extension Handle
5. Altitude Band Switch
6. Bailout Switch
7. Landing Gear Audio Warning Cutoff Button
8. Idle Thrust Control Switch
9. AFCS Transfer Controls
10. Master Electrical Power Switch
11. Radar IR Control Panel
12. Oxygen Control Panel
13. ILS Channel Selector Panel
14. Armament Control Panel
15. Throttle Quadrant
16. Throttle
17. UHF Control Panel
18. Takeoff Trim Light
19. Fuel Control Switch
20. Rudder Trim Switch
21. IFF Control Panel
22. Mask Defog Rheostat
23. Anti-G Suit Test Button
24. MA-1 Power Control Panel
25. Fuel Control Panel
26. MA-1 Test Panel
27. Cabin Air Selector Handle
28. Intercom Volume Control Panel
29. Variable Ramp Switch
30. Pitch G Limit Test Switch
31. UHF/TACAN Transfer Controls
32. Armament Recycle Button
33. Cockpit Left Fuse/Circuit Breaker Panel
34. Ram Air Turbine (RAT) Handle
35. Pressure Suit Control Handle
36. Reset/MBL Switch
37. Cockpit No-Fog and Ventilated Suit Switch
38. Landing and Taxi Light Switch
39. AIR-2A Arm/Safe/Monitor Power Circuit Breaker

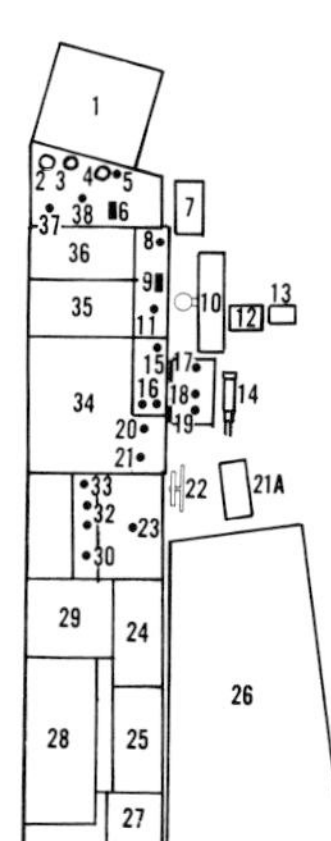

RIGHT CONSOLE

1. Master Warning Light Panel
2. Primary Hydraulic System Pressure Gage
3. Secondary Hydraulic System Pressure Gage
4. Oil Pressure Gage
5. IFF Caution Light
6. Generator Switch
7. No. 3 Fuel Tank Switch
8. Canopy Switch
9. Data Link Antenna Switch
10. Canopy Latch Handle
11. Warning Lights Test Button
12. ATG Switch
13. Map Reading Light Switch
14. Map Reading Light
15. EGT Spread Button
16. Windshield Anti-Icing, Antifog Switches
17. Emergency AC Generator Switch
18. Blank
19. TACAN-ADF Selector Switch
20. Engine Anti-Ice Warning Test Button
21. Rain Removal Switch
21A. Altitude Warning Selector
22. Ejection Seat Ground Safety Pin Stowage
23. Cabin Temperature Control Knob
24. Compass System Controller
25. Air Refueling Panel
26. Cockpit Right Fuse/Circuit Breaker Panel
27. MATTS Switch
28. Map and Data Case
29. Lighting Control Powerstats
30. Surface and Engine Anti-Icing Switch
31. Lighting Control Panel
32. Canopy Antifog Switches
33. Pitot Heat Switch
34. Data Link Control Panel
35. Auto-Navigation Homing Point Selector
36. TACAN Control Panel
37. Refrigeration Unit Switch
38. Cabin Air Selector Switch

18. TACAN Adjust Screw
19. Command Altitude Switch

BEFORE T.O. 1F-106-1157 (DME)

24. Compass Control Panel

BEFORE T.O. 1F-106-1181 (AHRG)

F-106B cockpit
-typical-aft

CONVENTIONAL INSTRUMENT DISPLAY

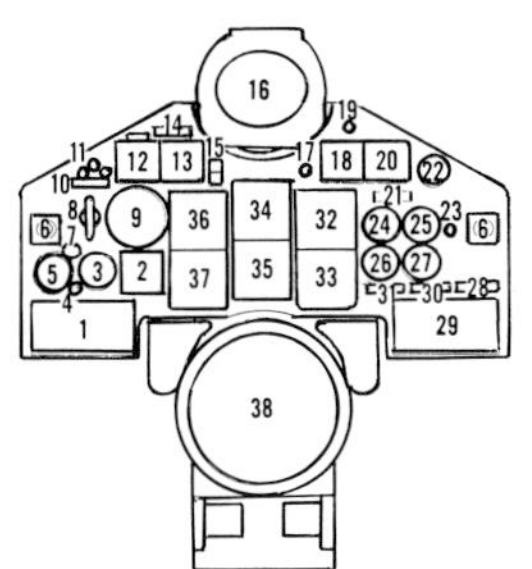

BEFORE T.O. 1F-106-1157 (DME)

4. TACAN Range Indicator Light
5. UHF Remote Indicator
9. Mach Indicator
32. Attitude Indicator
34. Course Indicator
35. Approach Horizon
37. Altimeter

Courtesy of the U.S.A.F.

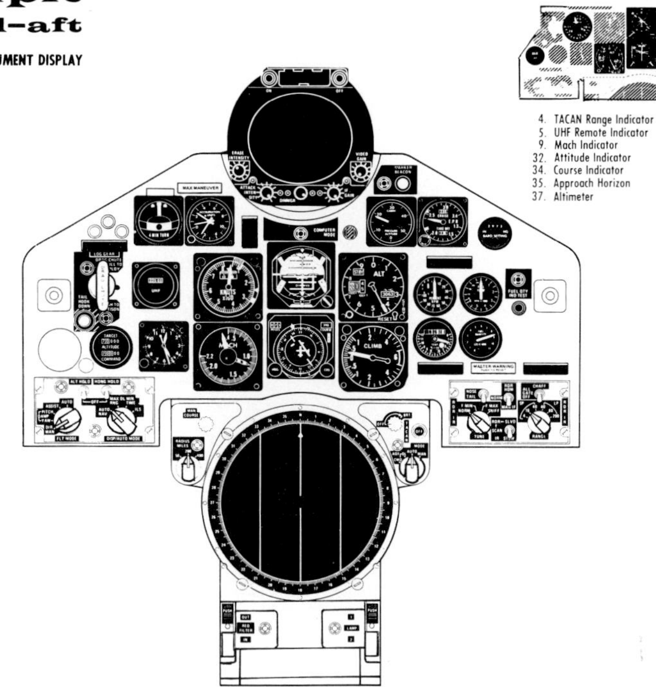

INSTRUMENT PANEL

1. Flight Modes Panel
2. Clock
3. Command and Target Altitude Indicator
4. Blank
5. Blank
6. Cabin Air Diffuser Head
7. Tail Hook Down Button and Light
8. Drag Chute Handle
9. UHF Remote Indicator
10. Landing Gear Warning Light
11. Landing Gear Position Lights
12. Turn-and-Slip Indicator
13. Accelerometer
14. Maximum Maneuver Warning Light
15. Bailout Warning Lights
16. Radar Scope
17. Computer Mode Indicator
18. Cabin Pressure Altitude Gage
19. Marker Beacon Light
20. Engine Pressure Ratio Gage
21. Engine Fire Warning Light
22. Barometer Setting Control
23. Fuel Quality Gage Test Button
24. Tachometer
25. Fuel Flow Indicator
26. Exhaust Gas Temperature Gage
27. Fuel Quantity Gage
28. Hydraulic Pressure-Low Warning Light
29. Radar/IR Selector Panel
30. Master Warning Light
31. Canopy Unlocked Warning Light
32. Altimeter
33. Vertical Velocity Indicator
34. Attitude Director Indicator
35. Horizontal Situation Indicator
36. Airspeed-Angle of Attack Indicator
37. Mach Indicator
38. Tactical Situation Display (TSD)

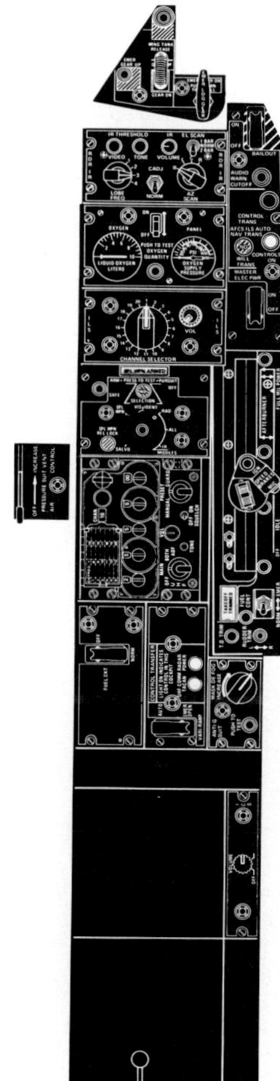

LEFT CONSOLE

1. Landing Gear Emergency Up Button
2. Landing Gear Handle
3. External Wing Tanks Release Button
4. Landing Gear Emergency Extension Handle
5. Bailout Switch
6. Radar/IR Control Panel
7. Landing Gear Audio Warning Cutoff Button
8. Oxygen Control Panel
9. AFCS Transfer Controls
10. ILS Channel Selector Panel
11. Master Electrical Power Switch
12. Armament Control Monitor Panel
13. Throttle Quadrant
14. Throttle
15. UHF Control Panel
16. Takeoff Trim Light
17. Fuel Control Switch
18. Rudder Trim Switch
19. UHF/TACAN Transfer Controls
20. Mask Defog Rheostat
21. Variable Ramp Switch
22. Anti-G Suit Test Button
23. Intercom Volume Control Panel
24. Cabin Air Selector Handle
25. Fuel Shutoff Switch
26. Pressure Suit Control Handle

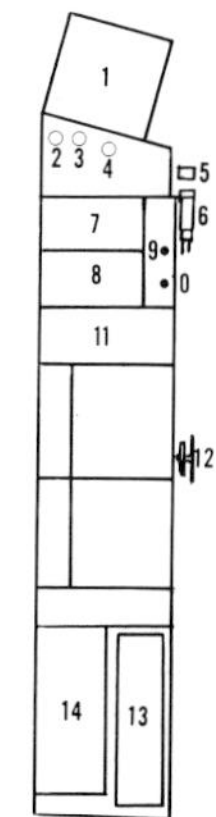

RIGHT CONSOLE

1. Master Warning Light Panel
2. Primary Hydraulic System Pressure Gage
3. Secondary Hydraulic System Pressure Gage
4. Oil Pressure Gage
5. Map Reading Light Switch
6. Map Reading Light
7. TACAN Control Panel
8. Auto-Navigation Homing Point Selector
9. Warning Lights Dimmer Switch
10. Warning Lights Test Button
11. Lighting Control Powerstats
12. Ejection Seat Ground Safety Pin Stowage
13. Cockpit Right Fuse/Circuit Breaker Panel
14. Map and Data Case

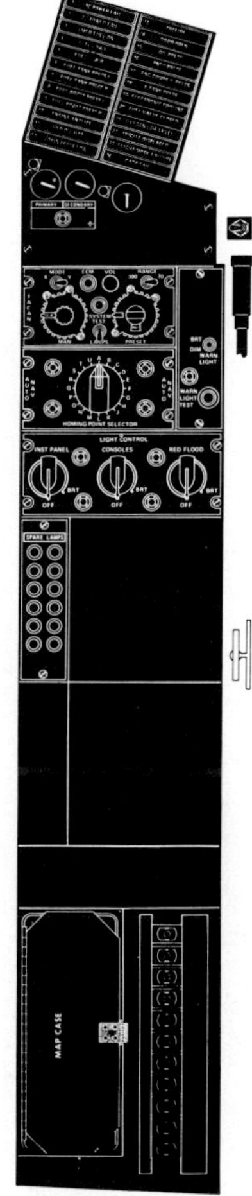

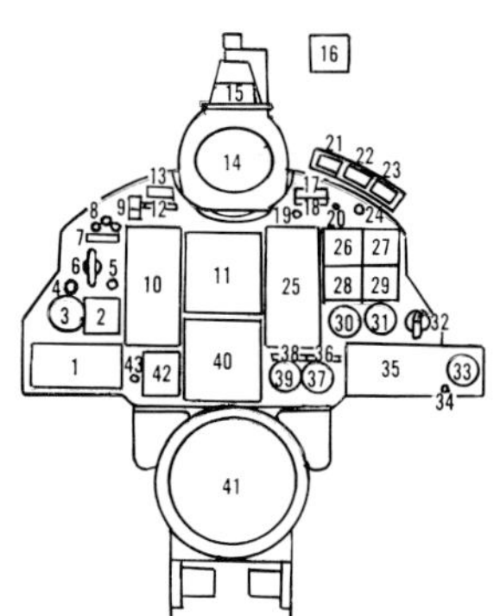

F-106B cockpit
-typical-forward

INTEGRATED FLIGHT INSTRUMENT SYSTEM

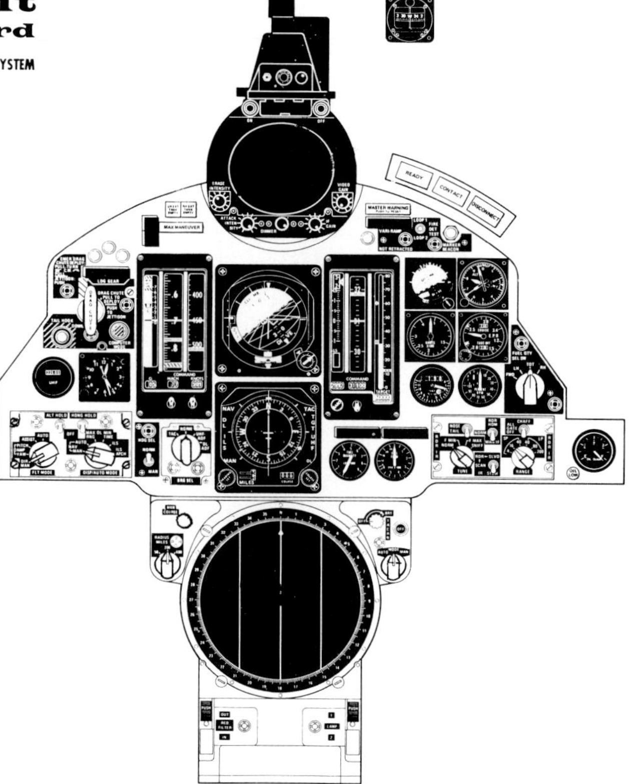

Courtesy of the U.S.A.F.

INSTRUMENT PANEL

1. Flight Modes Panel
2. Clock
3. UHF Remote Indicator
4. Tail Hook Down Button and Light
5. Computer Mode Indicator
6. Drag Chute Handle
7. Landing Gear Warning Light
8. Landing Gear Position Lights
9. Bailout Warning Lights
10. Airspeed-Mach Indicator (AMI)
11. Attitude Director Indicator (ADI)
12. Maximum Maneuver Warning Light
13. External Tank Empty Lights
14. Radar Scope
15. Radar Scope Recorder
16. Standby Compass
17. Master Warning Light
18. Engine Fire Warning Light
19. Variable Ramp Warning Light
20. Engine Fire Warning Test Switch
21. Air Refueling Ready Light
22. Air Refueling Contact Light
23. Air Refueling Disconnect Light
24. Marker Beacon Light
25. Altitude-Vertical Velocity Indicator (AVVI)
26. Standby Attitude Indicator
27. Standby Altimeter
28. Standby Airspeed Indicator
29. Engine Pressure Ratio Gage
30. Exhaust Gas Temperature Gage
31. Tachometer
32. Fuel Quantity Gage Selector Switch
33. Nucleonic Oil Quantity Gage
34. Oil Quantity-Low Light
35. Radar/IR Selector Panel
36. Hydraulic Pressure-Low Warning Light
37. Fuel Flow Indicator
38. Canopy Unlocked Warning Light
39. Fuel Quantity Gage
40. Horizontal Situation Indicator (HSI)
41. Tactical Situation Display (TSD)
42. Bearing Selector Switch
43. Heading Selector Switch

LEFT CONSOLE

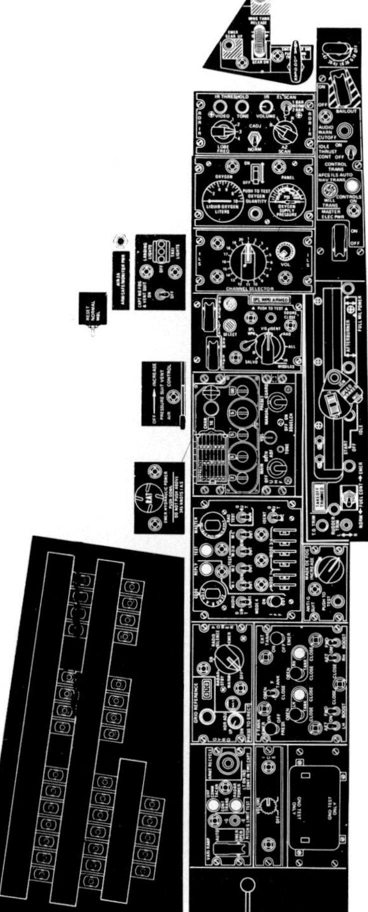

1. Landing Gear Emergency Up Button
2. Landing Gear Handle
3. External Wing Tanks Release Button
4. Landing Gear Emergency Extension Handle
5. Altitude Band Switch
6. Bailout Switch
7. Landing Gear Audio Warning Cutoff Button
8. Idle Thrust Control Switch
9. AFCS Transfer Controls
10. Master Electrical Power Switch
11. Radar/IR Control Panel
12. Oxygen Control Panel
13. ILS Channel Selector Panel
14. Armament Control Panel
15. Throttle Quadrant
16. Throttle
17. UHF Control Panel
18. Takeoff Trim Light
19. Fuel Control Switch
20. Rudder Trim Switch
21. IFF Control Panel
22. Mask Defog Rheostat
23. Anti-G Suit Test Button
24. MA-1 Power Control Panel
25. Fuel Control Panel
26. MA-1 Test Panel
27. Cabin Air Selector Handle
28. Intercom Volume Control Panel
29. Variable Ramp Switch
30. Pitch G Limit Test Switch
31. UHF/TACAN Transfer Controls
32. Armament Recycle Button
33. Cockpit Left Fuse/Circuit Breaker Panel
34. Ram Air Turbine (RAT) Handle
35. Pressure Suit Control Handle
36. Reset/MBL Switch
37. Cockpit No-Fog and Ventilated Suit Switch
38. Landing and Taxi Light Switch
39. AIR-2A Arm/Safe/Monitor Power

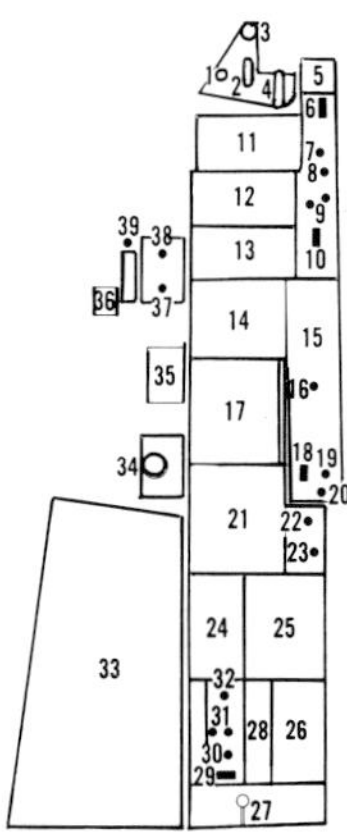

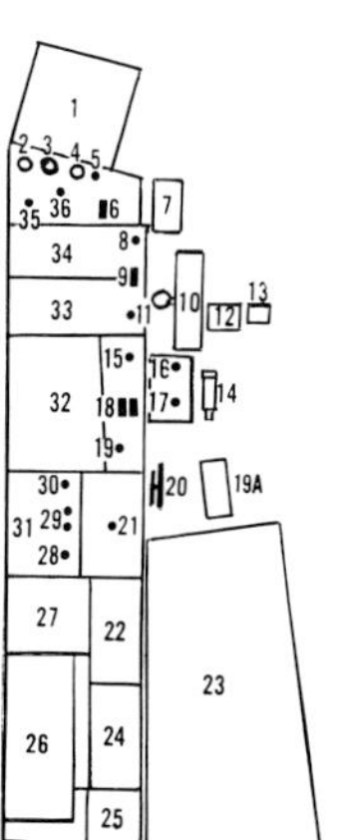

RIGHT CONSOLE

1. Master Warning Light Panel
2. Primary Hydraulic System Pressure Gage
3. Secondary Hydraulic System Pressure Gage
4. Oil Pressure Gage
5. IFF Caution Light
6. Generator Switch
7. No. 3 Fuel Tank Switch
8. Canopy Switch
9. Data Link Antenna Switch
10. Canopy Latch Handle
11. Warning Lights Test Button
12. ATG Switch
13. Map Reading Light Switch
14. Map Reading Light
15. Engine Anti-Ice Warning Test Button
16. Emergency AC Generator Switch
17. EGT Spread Button
18. Windshield Anti-Icing, Antifog Switches
19. Rain Removal Switch

19A. Altitude Warning Selector

20. Ejection Seat Ground Safety Pin Stowage
21. Cabin Temperature Control Knob
22. Compass System Controller
23. Cockpit Right Fuse Circuit Breaker Panel
24. Air Refueling Panel
25. MATTS Switch
26. Map and Data Case
27. Lighting Control Powerstats
28. Surface and Engine Anti-Icing Switch
29. Canopy Antifog Switches
30. Pitot Heat Switch
31. Lighting Control Panel
32. Data Link Control Panel
33. Auto-Navigation Homing Point Selector
34. TACAN Control Panel
35. Refrigeration Unit Switch
36. Cabin Air Selector Switch

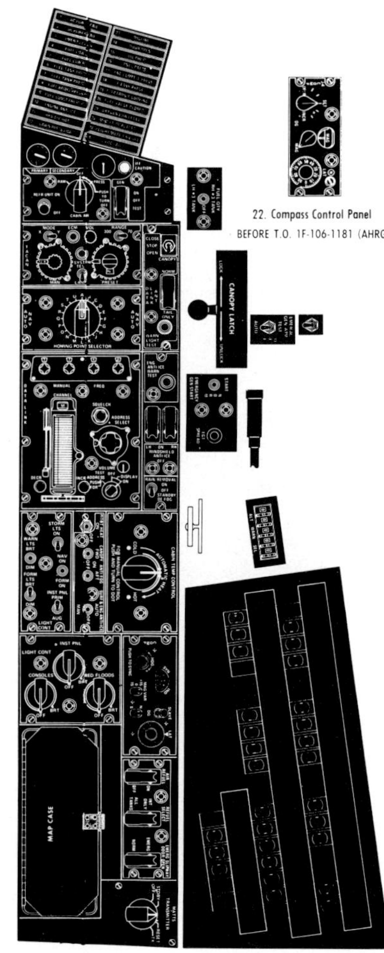

22. Compass Control Panel BEFORE T.O. 1F-106-1181 (AHRG)

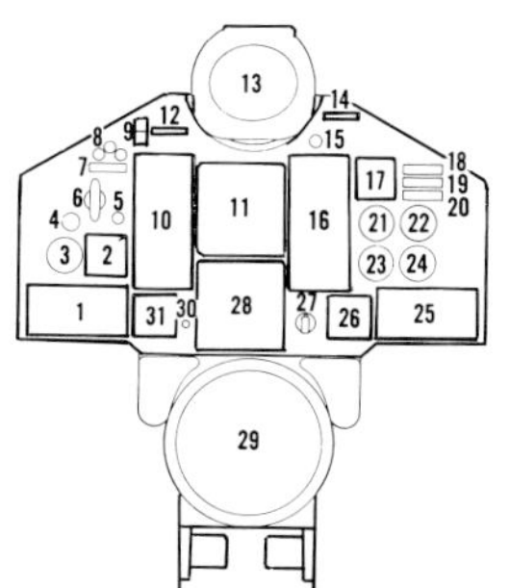

F-106B cockpit -typical-aft

INTEGRATED FLIGHT INSTRUMENT SYSTEM

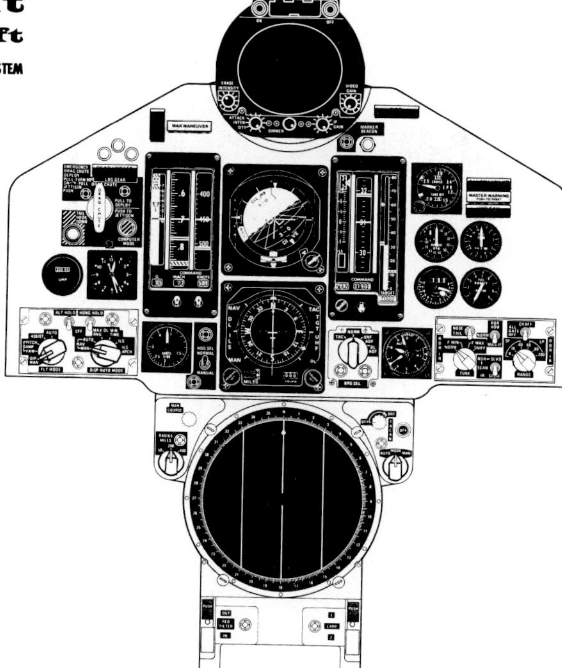

Courtesy of the U.S.A.F.

INSTRUMENT PANEL

1. Flight Modes Panel
2. Clock
3. UHF Remote Indicator
4. Tail Hook Down Button and Light
5. Computer Mode Indicator
6. Drag Chute Handle
7. Landing Gear Warning Light
8. Landing Gear Position Lights
9. Bailout Warning Lights
10. Airspeed-Mach Indicator (AMI)
11. Attitude Director Indicator (ADI)
12. Maximum Maneuver Warning Light
13. Radar Scope
14. Engine Fire Warning Light
15. Marker Beacon Light
16. Altitude-Vertical Velocity Indicator (AVVI)
17. Engine Pressure Ratio Gage
18. Canopy Unlocked Warning Light
19. Master Warning Light
20. Hydraulic Pressure-Low Warning Light
21. Tachometer
22. Fuel Flow Indicator
23. Exhaust Gas Temperature Gage
24. Fuel Quantity Gage
25. Radar/IR Selector Panel
26. Standby Altimeter
27. Bearing Selector Switch
28. Horizontal Situation Indicator (HSI)
29. Tactical Situation Display (TSD)
30. Heading Selector Switch
31. Standby Airspeed Indicator

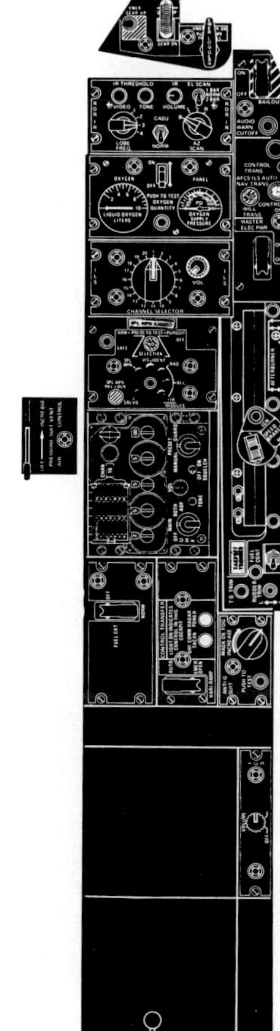

LEFT CONSOLE

1. Landing Gear Emergency Up Button
2. Landing Gear Handle
3. External Wing Tanks Release Button
4. Landing Gear Emergency Extension Handle
5. Bailout Switch
6. Radar/IR Control Panel
7. Landing Gear Audio Warning Cutoff Button
8. Oxygen Control Panel
9. AFCS Transfer Controls
10. ILS Channel Selector Panel
11. Master Electrical Power Switch
12. Armament Control Monitor Panel
13. Throttle Quadrant
14. Throttle
15. UHF Control Panel
16. Takeoff Trim Light
17. Fuel Control Switch
18. Rudder Trim Switch
19. UHF/TACAN Transfer Controls
20. Mask Defog Rheostat
21. Variable Ramp Switch
22. Anti-G Suit Test Button
23. Intercom Volume Control Panel
24. Cabin Air Selector Handle
25. Fuel Shutoff Switch
26. Pressure Suit Control Handle

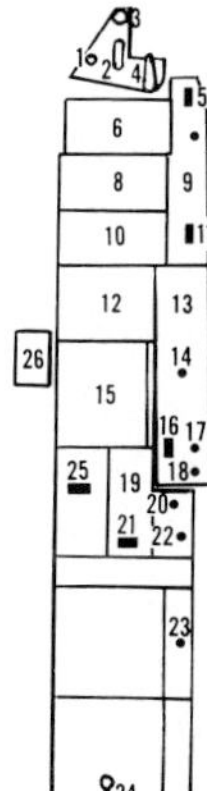

RIGHT CONSOLE

1. Master Warning Light Panel
2. Primary Hydraulic System Pressure Gage
3. Secondary Hydraulic System Pressure Gage
4. Oil Pressure Gage
5. Map Reading Light Switch
6. Map Reading Light
7. TACAN Control Panel
8. Auto-Navigation Homing Point Selector
9. Warning Lights Dimmer Switch
10. Warning Lights Test Button
11. Lighting Control Powerstats
12. Ejection Seat Ground Safety Pin Stowage
13. Cockpit Right Fuse/Circuit Breaker Panel
14. Map and Data Case

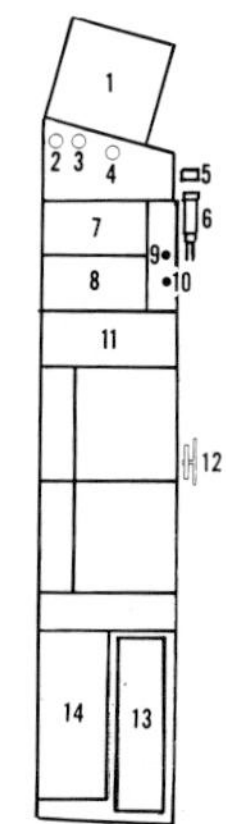

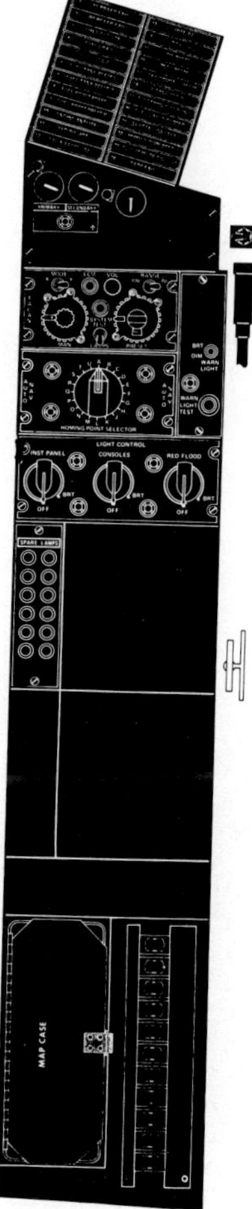

F-106B WEAPONS BAY DETAIL

Above: View looking aft at the right weapons bay doors. Note the differences from those used on the F-106A as shown on page 42.

Left: Because of the addition of the second cockpit, the lower electronics bay on the F-106B had to be located between the forward two missiles. This caused the differences in the doors, and eliminated the use of the crossbridge structure as seen on the F-106A.

Below left and right: Two views showing the separate launch rails for the forward two missiles on an F-106B.

Looking forward in the weapons bay of an F-106B.

View looking aft showing the rear missile launch rails in the retracted position. Also note the bay door actuator arms.

Rear missile launch rails in an F-106B shown in the retracted position. The AIR-2 Genie rocket fits on the pallet at the center of the photo.

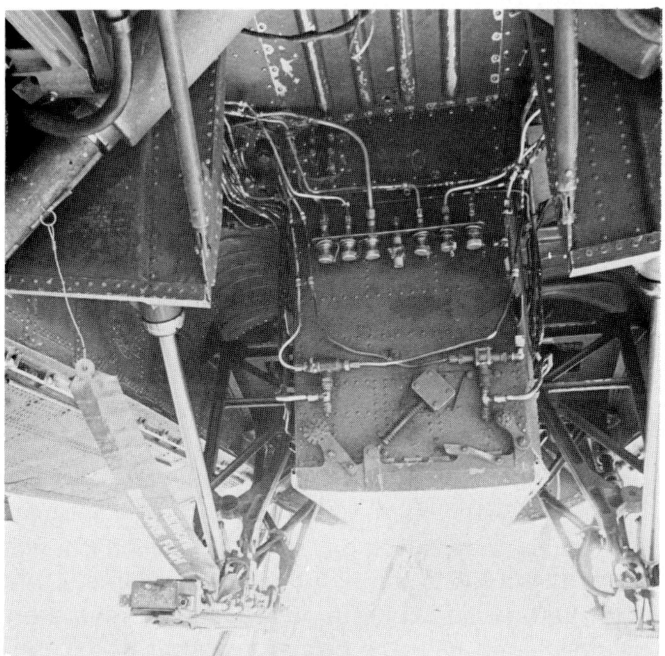

Rear view of the left front missile launch rail in the extended position.

MODELER'S SECTION

PRODUCT REVIEW POLICY. In each of our publications we will try to review kits and decals that are available to the scale modeler. We hope to be able to review every currently available kit that is useable by the scale modeler. Kits produced in the past that are no longer generally available, and those more intended to be toys than accurate scale models will not usually be covered. Additionally, we do not intend to give a complete step-by-step, correction-by-correction account of how to build each kit. Instead we intend to give a brief description of what is available to the modeler, and point out some of the good and not-so-good points of each kit or product. In this way we hope to give an overall picture of what the modeler has readily available for his use in building the particular aircraft involved.

KIT REVIEWS

OFF-SCALE KIT

REVELL 1/71st (?) SCALE F-106A, KIT NUMBER H-159

The instruction sheet for this kit claims that the model is 1/71st scale, but it really is not any scale. The wingspan works out to about 1/66th scale, while the length is about 1/64th scale. But the fuselage cross-section is about right for 1/72nd scale, as is the canopy. But since this kit is copyrighted in 1961, Revell was obviously working with sketchy information when they did the research for this kit. To their credit, Revell tried to incorporate some of the early changes to the Dart in their model. For example, the model features the slots in the wing as opposed to the wing fences. The longer afterburner can is present, as is the vent in the spine.

Usually we would not review a kit such as this one since it is not generally available anymore, but since there are so few kits of the F-106 that are available, and since we came across one at the 1982 IPMS National Convention, it seemed almost irresistable to give this old kit a try. (A word of caution is due here. Carefully inspect any kit purchased from those "collectors" who sell their kits at conventions and elsewhere. When we got our kit home, we found that the fuselage had been glued together, then pried apart, breaking it badly. There was no notice of this on the box, yet the seller represented it as a kit just as it was boxed from Revell. Not all are this dishonest, but some deserve close watching.)

The model is typical of kits from the late 50's and early 60's with major markings engraved into the plastic. Surface scribing is recessed, which is a nice feature, and rivals some of the best available today. The ailerons are separate pieces, and can therefore be positioned with the proper droop at which they rest when the aircraft is on the ground. Likewise, the rudder is a separate piece, and can be positioned off center. The speed brake and canopy are hinged, and can be built to open and close. But the hinges are anything but scale.

The model lacks a pitot tube, and one should be added. The blade antenna, located just behind the nose gear, and present on the first F-106s, is present. The landing gear is bulky and anything but accurate. The main gear wells are incomplete, and when represented, the wells are merely shallow recessed areas in the wing and nose. They are not deep enough to accommodate the gear. However, the wheels fairly accurately represent the earlier wheels with the spoked pattern.

The cockpit consists of a seat and a pilot figure. No console or instrument panel is present.

The markings are for the third F-106, 56-453, and it sports the bright orange areas on the nose, tail, and wings. On our sample, we changed the number to 56-455, and based the scheme on a photo of that F-106 taken in flight. The decals include stenciling and even the position lights under the wings. Compared to other kits of its day, this kit's decals are quite extensive.

The true value of this kit really lies in the nostalgic value placed on it by collectors. It really isn't an accurate scale model, but is a representation of the F-106. Building it was a nostalgic "trip" to the days when model kits lacked the detail and excellent fit of most of today's offerings. It's an interesting kit, but not very authentic.

The old Revell kit built as the fifth F-106A.

1/72nd SCALE KIT

MINICRAFT/HASEGAWA 1/72nd SCALE F-106A, KIT NUMBERS 1054 and JS-054.

This kit dates back to the 60's and does show its age. Fit is very poor and requires a great deal of filling and sanding. But the outline is generally accurate, and it can be built into a most attractive model. It has been released three different times. Between the first and second releases only the box art was changed, and even the decals remained the same. In the third release, new decals were included and featured the colorful "Freedom Bird" bicentennial markings. But updates were not made to the molds, so those that were made to the actual aircraft after the first release of the kit are missing from the model and will have to be added as appropriate by the modeler. Additionally, one noticeable error must be corrected as a minimum. Hasegawa represented the wing slots as small boundary layer fences. These should be removed and replaced with slots carefully cut in the wing with a razor saw. Also, the air intakes require a bit of reshaping. Refer to the drawings and photos in this book for the proper shape.

Since this is the only available kit of the F-106 in 1/72nd scale, the modeler has no other choice if he is to model the Dart in this popular scale. Therefore, as a departure from our regular review, we are presenting below a narrative on how to build the kit as accurately as possible. With relatively easy-to-make changes, the modeler can build the kit into any F-106 from the earliest prototype to the latest configuration of the Dart now in use.

The kit has the production IR hump on the nose, the vent on the spine, and a tail hook. Supersonic late model fuel tanks are included, and the main gear wheels more closely represent the later style wheels. So that is where you begin. The first step is to get good *photographic* reference for the specific aircraft you intend to build. A prototype aircraft will have wing fences, and Minicraft/Hasegawa's F-102 kit can be used as a reference for making these from thin plastic card. The blade antenna behind the nose gear will also have to be added from plastic card, and the vent on the spine may or may not need to be filled depending on the aircraft and/or time period involved. The IR sensor hump should be sanded smooth, and the main gear wheels from the F-102 kit should be used. The tail hook should be deleted with the locator hole filled and the hook guards removed. The engine nozzle will have to be shortened. With these modifications made, a good prototype will result.

To build an early F-106, the IR sensor may or may not be present, and again the early style wheels from an F-102 kit will be needed. Likewise, the older subsonic 230 gallon fuel tanks can be obtained from the F-102 kits.

As time progressed, updates and modifications were made to the F-106 as outlined in this book. The first step is for the modeler to determine if the IR fairing or hook is needed, then use or delete as appropriate. Use the wheels from the F-102 kit for the old style, and the F-106 kit for the late style. Choose the appropriate tanks from the F-102 kit or F-106 kit. If the tanks from the F-106 kit are used, the point at the trailing edge should be filed down a bit. It is blunt as shown on page 45. Other updates must be made from scratch by the modeler, and this is where the detail photos in this book come in. The optical sight is not included in the kit, and can be made from sprue. The data link antennas, located next to the tail hook, can be cut from thin plastic card. The upper rotating beacon can be made from a short piece of sprue, and the refueling receptacle can be made with the addition of thin plastic strips and some cutting and shaping.

The kit provides the older style canopy with the top framing. If the later blown canopy is required, it may be ordered from the Bare-Metal Foil Company. Send a SASE to 19419 Ingram, Livonia, Michigan 48152 for details.

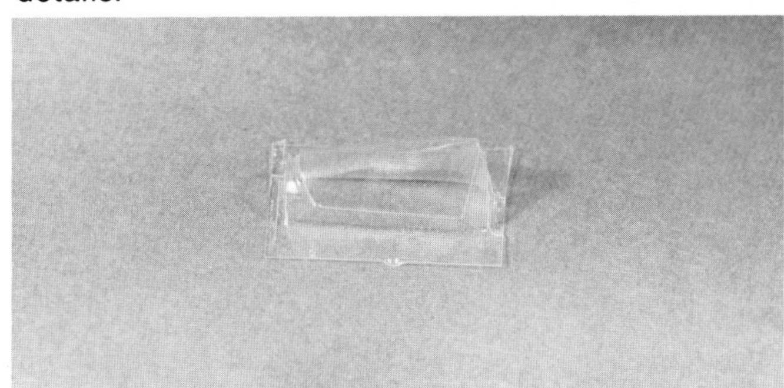

Bare Metal produces a very nice clear "bubble" canopy for the F-106A in 1/72nd scale.

The last modification to consider is the gun. This fairing will have to be made from scratch, or the fairing from an Airfix/MPC F-4 Phantom kit. The fairing required is the one for the gun for the F-4E provided in the kit. However, the rear of the fairing will have to be filled in and shaped. The best way to proceed is to build the weapons bay closed, and fit the fairing over the doors in the proper location.

With all the updates added or deleted as required, the model still needs some detailing and correcting. Areas requiring attention include the cockpit, wheel wells, and weapons bay.

Minicraft/Hasegawa kit built in markings of the 27th FIS.

The cockpit comes with a floor/console combination that has a locator hole for the seat. No instrument panel or control column is provided, and the pilot figure leaves a bit to be desired. The two humps on the consoles should be removed, and the tops of the consoles should be covered with plastic card. Cut the plastic card a little larger than the tops of the consoles, then glue them to the consoles using tube glue. Before the glue sets, place the two fuselage halves around the floor/console part, and press together, but do not glue. This will cause the pieces of plastic card to align flush with the sides of the fuselage cockpit walls. Remove the fuselage halves, and wait for the glue to dry on the consoles. After it has dried, trim the pieces at the front, aft, and inside edges to match the sides of the consoles on the kit part. This insures a perfect fit on all sides, and a flush fit against the inner fuselage. Next add a throttle quadrant, then a Y-shaped control yoke. The instrument panel can be made from plastic card. The cockpit is now ready for detailing. The kit seat forms the basis for a good ejection seat with some detailing, and the area behind the seat, to include the canopy raising and lowering mechanism, should be detailed.

The nose gear well is deep enough to use from the kit, with detailing a relatively simple chore. Modelers will want to make sure that the light on the front of the nose gear door is present.

The main gear wells require a lot more work. The outer portions (under the wings) need to have the sides added so that there is no gap between the upper and lower wing pieces. This can rather easily be done with thin plastic card. A second actuating strut should be connected to each main strut. The inner portion of the wells is totally wrong and needs to be replaced. All that is provided in the kit is a vertical piece at the aircraft centerline to which several ribs are attached. This should all be cut away and replaced. A rounded area, representing the lower portion of the engine casing, should be added, and can be vacu-formed or cut from an appropriately sized fuel tank or bomb from the parts box. The engine part from a Monogram B-52 is excellent. The center piece, to which the doors are attached, is then required from plastic stock. Now all that is left is the detailing which can be accomplished with fine wire and scraps of plastic.

The main gear doors also require some detailing. There are rods on the forward and aft edges of the inner doors, and these can be added using sprue or appropriate parts from the parts box. (See page 19 for details.) The outer doors have braces running between them and the strut, and these must be added from scratch. Lights should also be added to these outer doors. (See pages 17 and 18 for details.)

The last area for consideration is the weapons bay. The first point here is to decide if it is to be opened or closed. If it is to be closed *do not* try to glue the four kit doors over the bay. Instead, cut a thick piece of plastic card to match the size of the hole in the bottom of the fuselage. Glue this in place, and then sand the outer edges to the rounded shape of the fuselage. Next fill with putty and sand smooth. Then scribe in the lines of the bay and the door very carefully. Lastly add two stripes of thin stretched sprue to represent the piano hinges between the outer doors and the fuselage.

If the bay is to be open, all kinds of problems arise. First, it is too shallow, and correcting this can be a monumental task. If you use the inner piece as it comes in the kit (part 27), the problem will not be too noticeable unless you pick the model up and turn it over to look at it. If part 27 is used, the rails and crossbridge support must be used in the extended position since there is not enough room for them to fit into the bay in the retracted position. It should be noted that the interior of the bay has no real detailing and no AIR-2 Genie rocket or associated launch rail. If the modeler wants to add the proper depth to the weapons bay, he should get some plastic card, a few hundred spare hours, and open this book to pages 42 and 43.

Two other problems exist with the weapons bay. First, be careful to get the doors aligned correctly. The instructions do not show this properly, but the photo on page 42 does. It will help to thin down the doors and shorten the hinges a little. The second problem area is the missiles themselves. The two rear missiles (supposedly the IR missiles but called "Nuclear Falcons" in the instructions) have probes on the noses. These represent the old AIM-4E which has been out of the inventory for many years. The probes should be removed. Further, none of the missiles have the gap between the wing and the control fin, nor do they have the hinge between the wing and fin. Therefore the missiles have a "toyish" appearance, and just don't look like the real thing. However, the remedy is fairly simple. Carefully cut the gap between the wing and the fin using a razor saw, then add the hinges using stretched sprue or very thin strips of plastic card. This worked quite well on our sample models, and greatly added to the realism of the finished product.

While the kit does not compare to the latest releases from Hasegawa, it remains a basically accurate kit that can be made into an authentic and realistic model. By modeling one of the beautiful color schemes that were applied to the Delta Dart, the model can easily become one of the most attractive in any collection. We recommend this kit.

1/48th SCALE KIT

MONOGRAM 1/48th SCALE F-106A, KIT NUMBER 5809

Monogram's 1/48th scale kits of the Century Series fighters just keep getting better and better! Following their beautiful F-105G, this kit tops them all for features and detailing. While the production kit was not released before press time, Monogram was kind enough to provide us with a test shot to examine. The decals and instruction sheet were not available, however the decals will be for the Red Bulls of the 87th FIS. The featured aircraft is the commander's bird, and is named the "City of Marquette." The markings feature commander's stripes on the fuselage, and a bicentennial "pretzel." Detail & Scale was pleased to provide the reference for these markings.

The kit is really beautiful with detailing everywhere. The cockpit has a very nice ejection seat, and an excellent "Y" shaped control column. The instrument panel and consoles are nicely detailed with instruments and the appropriate knobs and switches. The only thing missing is the throttle, which has been the one shortcoming of all of Monogram's 1/48th scale fighters.

To cover the cockpit, you get both the older canopy and the newer blown canopy. Either may be shown in the open position, with detailing provided behind the seat where the raising and lowering mechanism is.

The wheel wells and landing gear are all delicately detailed with crisply molded features. Antennas are also nicely done with the forward blade and aft data link antennas provided. The tail hook, refueling receptacle, IR sensor hump, late style wheels and supersonic fuel tanks are all provided, and indicate that the model is based on a recent F-106.

The most eye-catching and best detailed area is the weapons bay. It is beautifully done, and is as detailed as the state of the art in injection molding will allow. Launch rails, hydraulic lines, and a crossbridge assembly are all present to make this an amazing replica of the real thing. Both Genie and Falcon missiles are provided, and, in the case of the Falcons, may be mounted on their rails in the extended or retracted position. Two sets of doors are provided for the model, one set molded in the correct open position in pairs, and the second being molded as all four doors in the closed position. However, it would seem a crime to close this highly detailed area off and not use it to enhance the beauty and realism of the model.

Other features include a RAT that may be shown in the open position, ailerons molded in the proper drooped angle, and speed brakes that may be displayed open or closed.

There are many features to rave about, but suffice it to say that it is one of the best kits ever marketed. Even those modelers who have always built in 1/72nd scale, and avoided the larger scales, owe it to themselves to experience the challenge and reward of building this excellent and highly detailed kit. We highly recommend this kit as one of the best ever produced.

Monogram's beautiful F-106 model in 1/48th scale. This is one of the best 1/48th scale kits ever made. (Monogram)

1/32nd SCALE KIT

COMBAT MODELS 1/32nd SCALE F-106A, KIT NUMBER 32-043

Combat Models has just released their new F-106A in 1/32nd scale. Detail & Scale was lucky enough to receive the first kit off of the molds, and it really looks nice. This vacu-formed model features a clear canopy (of the older framed variety) and metal landing gear struts. Surface scribing is very nice, and in general, this is one of the best vacu-formed models we have seen. Since the kit arrived less than a week before the deadline of this book, we did not have time to build it, but it has captivated our interest, and will soon be a completed model on the shelf.

The kit comes with fuselage halves, top and bottom pieces for each wing, separate speed brakes, supersonic style fuel tanks, left and right sides for the vertical tail, air intakes, afterburner can halves, outer main landing gear doors, and main gear wheels molded into the white plastic. With the exception of the main gear wheels, all parts are beautifully formed and accurate. The main gear wheels look nice, but don't really look like either style of wheel used on the F-106, so the modeler will probably want to rework them. The nose wheels are among the metal parts supplied with the landing gear struts.

A large size sheet of thick plastic card is supplied for making such things as pylons, hinge fairings, the nose gear door, and other small flat parts.

The kit itself features the IR sensor hump, the vent on the spine (which could easily be converted to a refueling receptacle), the wing slots, framed style canopy, and supersonic tanks as mentioned earlier. The roomy cockpit area just waits to be detailed, and the modeler with a few years to spare might even try to open the weapons bay!

This kit is really a nice one, and is well worth the price and effort it would take to turn it into a true masterpiece. For further information, send a SASE to, Combat Models, 1633 Marconi Road, Wall, New Jersey 07719. We recommend this kit.

DECAL SUMMARY

Note: It is impossible to completely review decals unless the reviewer has actually used the decals on a model to see how they fit. Additionally, markings on a given aircraft can be changed from time to time, so it is possible that the decals may be accurate for one point in time and not another. Therefore, this section is more of a listing of decals available than a review. Review comments are made only in regard to fit when we have actually used the decals or as to accuracy when the evidence clearly indicated an error.

KIT DECALS

OFF-SCALE KIT

Revell F-106A, Kit Number H-159: Provides markings for the third F-106A, 56-453. Markings are intended for the high visibility scheme with red/orange panels on the wings, tail and nose.

1/72nd SCALE KIT

Hasegawa F-106A, Kit Number JS-054, (early issue): Provides markings for two aircraft.

- F-106A-100-CO, 58-0779, 94th FIS, Selfridge AFB, Michigan
- F-106A-1-CO, 57-0246, 95th FIS, Dover AFB, Delaware

Minicraft/Hasegawa F-106A, Kit Number JS-054 (later issue): Contains markings for three aircraft.

- F-106A-125-CO, 59-0094, 87th FIS, "Red Bulls," K.I. Sawyer AFB, Michigan, aircraft name is "Bones Crusher"
- F-106A-125-CO, 59-0089, 87th FIS, "Red Bulls," K.I. Sawyer AFB, Michigan, aircraft name is "Lurch IV"
- F-106A-110-CO, 59-0057, 318th FIS, McChord AFB, Washington

Minicraft/Hasegawa F-106A, Kit Number 1054: Provides markings for an F-106A-100-CO, 58-0776, of the 318th FIS, McChord AFB, Washington. The aircraft has an elaborate bicentennial scheme, and is named the "Freedom Bird." These decals are more accurate in size an proportion to those provided on Microscale's sheet number 72-196.

AMT/Hasegawa F-106A, Kit Number A690: Contains markings for two F-106As.

- F-106A-100-CO, 58-0787, 94th FIS, Selfridge AFB, Michigan. "Hat in Ring" squadron
- F-106A-85-CO, 57-2480, 456th FIS, Castle AFB, California

Note: This kit is a reissue of the standard Minicraft/Hasegawa kit.

1/48th SCALE KIT

Monogram F-106A, Kit Number 5809: Provides markings for an F-106A-110-CO, 59-0053, 87th FIS, "Red Bulls," K.I. Sawyer AFB, Michigan. Aircraft name is "City of Marquette," and command stripes are on the fuselage.

DECALS SHEETS

1/72nd SCALE SHEETS

Aerodecal Sheet Number 22A: Provides markings for four aircraft.

- F-106A-100-CO, 58-0783, 2nd FIS, "The Horney Horses," Wurtsmith AFB, Michigan
- F-106A, 125-CO, 59-0094, 87th FIS, "Red Bulls," K.I. Sawyer AFB, Michigan. Aircraft name is "Bones Crusher."
- F-106A-1-CO, 57-0230, 87th FIS, "Red Bulls," K.I. Sawyer AFB, Michigan. Aircraft name is "Lurch IV," and is alternative markings for 59-0094 above.
- F-106A-1-CO, 57-2508, Air Defense Weapons Center, Tyndall AFB, Florida

Bare Metal Sheet Number 72-1: Provides markings for an F-106 from the 191st Fig, Michigan ANG, Selfridge AFB, Michigan. A number strip allows a choice of aircraft tail number. Aircraft include "Defiance" (58-0772), "Polish Kid" (58-0793), and an aircraft with an American Eagle with a flag (56-0465). Markings for a T-33 are also provided.

Bare Metal Sheet Number 72-6: Provides markings for an aircraft from the Massachusetts ANG. A number strip allows choice of tail number.

Bare Metal Sheet Number 72-7: Provides markings for an aircraft from the New Jersey ANG in that unit's present (1983) markings with stylized chevron and the words "New Jersey." A number strip allows a choice of aircraft tail number.

Note: If Bare Metal decals are not available in your area, send a SASE to Bare Metal Co., 19419 Ingram, Livonia, Michigan 48152 for ordering information.

Microscale Sheet Number 72-91: Provides markings for three aircraft.

- F-106A-125-CO, 59-0130, 84th FIS, Hamilton AFB, California
- F-106A-135-CO, 59-0145, 318th FIS, McChord AFB, Washington. Pilot's name is Major Schoensiegel
- F-106A-80-CO, 57-2467, 539th FIS, McGuire AFB, New Jersey

Microscale Sheet Number 72-92: Provides markings for four aircraft.

- F-106A-110-CO, 59-0044, 27th FIS, Loring AFB, Maine
- F-106B-65-CO, 58-0900, 27th FIS, Loring AFB, Maine. Cartoon caricature of a pranged plucked duck with a sign "Flying without feathers is not easy" is on the nose. Pilot name is Lt. Col. Stewart. Note - Requires conversion.
- F-106A-1-CO, 57-0236, 5th FIS, Minot AFB, North Dakota
- F-106A-120-CO, 59-0081, 49th FIS, "Green Eagles," Griffiss AFB, New York

Microscale Sheet Number 72-115: This sheet provides the markings for an aircraft from the 46th FIS, the 2nd FIS, and one of three aircraft from the 87th FIS.

- F-106A-70-CO, 57-2456, 46th FIS, Oxnard AFB, California
- F-106A-100-CO, 2nd FIS, "The Horney Horses," Wurtsmith AFB, Michigan
- One of three from the 87th FIS, "Red Bulls," K.I. Sawyer AFB, Michigan
  F-106A-125-CO, 59-0088, "Jack the Gripper "
  F-106A-125-CO, 59-0089, "Lurch IV "
  F-106A-125-CO, 59-0094, "Bones Crusher "

Microscale Sheet Number 72-196: Provides markings for five aircraft.

- F-106A-100-CO, 58-0776, 318th FIS, McChord AFB, Washington. This is the same bicentennial scheme as provided in the Minicraft/Hasegawa kit number 1054. These do not seem as accurately proportioned as those in the kit. Specifically, the red, white, and blue bands around the fuselage appear too wide.
- F-106A-135-CO, 59-0138, Air Defense Weapons Center, Tyndall, AFB, Florida

- F-106B-80-CO, 59-0165, Air Defense Weapons Center, Tyndall, AFB, Florida. These are alternate markings for 59-0138 above. This is the personal aircraft for General Daniel "Chappie" James, CINCNORAD. Note - Requires conversion.
- F-106A-130-CO, 59-0127, 48th FIS, Langley AFB, Virginia. Pilot's name is Captain Emerson. Note - The decal sheet is correct, however the instruction sheet fails to show the location of the Hughes Trophy award, and does not show the word "PILOT" before the pilot's name.
- F-106B-75-CO, 59-90157, 48th FIS, Langley AFB, Virginia. These markings provide an alternate for 59-0127 above.

Microscale Sheet Number 72-197: Provides markings for three aircraft.

- F-106A-1-CO, 57-0240, 177th FIS, New Jersey Air National Guard
- F-106A-85-CO, 57-2485, 186th FIS, Montana Air National Guard
- F-106A-100-CO, 58-0790, 194th FIS, California Air National Guard

Scalemaster Sheet Number SM-9: Provides markings for the colorful Florida ANG, bicentennial aircraft, 58-0760, named the "City of Jacksonville."

1/48th SCALE SHEETS

Bare Metal Sheet Number 48-5: Provides markings for an aircraft from the Michigan Air National Guard. Includes fifteen pilots' names and twelve crew chiefs' names plus a number strip for selection of aircraft tail number. Instructions tell which name goes with each number.

Bare Metal Sheet Number 48-6: Provides markings for an aircraft from the Massachusetts ANG. A number strip allows selection of aircraft tail number.

Bare Metal Sheet Number 48-7: Provides markings for an aircraft from the New Jersey ANG in that unit's present (1983) markings with stylized chevron and the words "New Jersey." A number strip allows choice of aircraft tail number.

Detail & Scale Sheet Number 0448, F-106s of the 48th FIS, Early and Late Markings: Provides markings for four aircraft from the 48th FIS, Langley AFB, Virginia.

- F-106A-130-CO, 59-0135, 1972 markings, pilot's name is Captain Roeder
- F-106A-130-CO, 59-0127, 1972 markings, pilot's name is Captain Emerson
- F-106A-130-CO, 59-0127, 1976 markings. This is the same as the aircraft above, but with new markings and the blown canopy.
- F-106A-1-CO, 56-0458, 1978 William Tell markings. Aircraft name is "Thunderchicken," and has a cartoon chicken on the nose. Pilot's name is Lt. Col. Bob Altwater.

Detail & Scale Sheet Number 0548, Active Air Force F-106A Delta Darts: Provides markings for three aircraft.

- F-106A-125-CO, 59-0103, Air Defense Weapons Center, October 1978. Red, White, and blue command stripes on fuselage
- F-106A-100-CO, 58-0780, 460th FIS, William Tell 1972 markings, Grand Forks AFB, North Dakota
- F-106A-100-CO, 58-0773, 94th FIS, "Hat in Ring" squadron

Fowler Sheet 48-16: Provides markings for two F-106As.

- F-106A-120-CO, 59-0082, 49th FIS, "Green Eagles," William Tell 1982, Command stripes on fuselage
- F-106A-105-CO, 59-0004, 318th FIS, commander's aircraft, McChord AFB, Washington

Fowler Sheet 48-17: At press time the aircraft for this sheet had not been determined.
For additional information on Fowler Decals, send a SASE to Fowler Aviation, P.O. Box 148, Sunnymead, California 92338.

Note: No Microscale sheets for the F-106A had been released in 1/48th scale at press time for this book. However, we believe Microscale will release several sheets for the Monogram 1/48th scale kit shortly after it becomes available. If Microscale follows their past practices, it is likely that these 1/48th scale sheets will contain many of the same markings found on their 1/72nd scale sheets listed above. However new markings may also be included.

REFERENCE LISTING

Listed here are references on the Delta Dart that should prove helpful in providing information and photographs of a different nature and format than what is presented in this publication. With each listing is a brief description of what the reference covers.

Note: While we usually only list references rather than review them, it should be noted that several of the references listed below have some serious errors concerning the F-106. One book contained a photograph of the inner main landing gear doors closed over the fuselage, and the caption stated that the photo was of the weapons bay. One magazine article included a drawing of an F-106A with the F-106B weapons bay and call-out pointing to the "Lower Aft Electronics Compartment" in the forward end of the bay. This is incorrect for an F-106A. No scale drawing could be found that was accurate. Usually, drawings measured out to one scale in span and another in length. Another problem area was found with drawings depicting color schemes. Schemes used early in the operational life of the F-106 were shown on drawings with the later modifications such as the refueling receptacle shown. Likewise, drawings of color schemes applied to aircraft with modifications were shown with those modifications missing. When researching markings for a specific F-106, such as for modeling purposes, great care should be used, and photographs, not drawings, should be consulted. With the many changes applied to the F-106 airframe during its operational life, most noticeably those shown on pages 10-14 of this book, the researcher should take care to determine which changes apply to the aircraft at the specific point in time he is researching.

BOOKS:

1. Carson, Don, and Lou Drendel, F-106 Delta Dart in Action, Squadron/Signal Publications, Carrollton, TX. 1974

Contains numerous photos of F-106s "in action," and an account of what it is like to fly the aircraft by Don Carson, a former F-106 pilot.

2. Holder, William G., Convair F-106, Aero Publishers, Fallbrook, CA. 1977

One of the books in the Aero series, this title covers the developmental history of the Delta Dart.

3. Kinzey, Bert, Colors & Markings of the F-106 Delta Dart, Aero Publishers, Fallbrook, CA. 1983

The first in a new series of publications due for release in the summer of 1983, this book covers the markings of every active Air Force and Air National Guard unit ever to fly the F-106. The Air Defense Weapons Center aircraft are covered as are those used by NASA. All markings are shown in photographs, not drawings, and 16 pages of full color are included.

4. F-106 Delta Dart, Koku Fan Special Number 128, Bunrindo Publications, Bunrin-Do, Japan, 1981

A pictorial booklet that is an excellent source for markings information. Japanese text.

5. F-102/F-106, Koku Fan Special Number 51, Bunrindo Publications, Bunrin-Do, Japan, 1974

Typical Koku Fan pictorial with some coverage of the F-106's markings through 1974. Japanese text.

MAGAZINE ARTICLES:

1. Peacock, Lindsay, "Convair F-106 Delta Dart," Scale Aircraft Modeling, Vol. 3, No. 7, April 1981, Page 297

Short article summarizing the development of the F-106. Includes numerous drawings showing markings.

2. Wogstad, James, and Phillip Friddell, "Convair F-106 Delta Dart," Replica in Scale, Vol. 1, No. 4, Summer 1973, Page 120

Brief article covering the F-106 with good general photographs and some details shown in drawings. Numerous drawings show F-106 markings.

3. "Warpaint - F-106 Delta Dart," Aviation News, Vol. 5, No. 22, April 1-14, 1977, Page 7

Several drawings show markings carried by F-106 aircraft.

4. Cartmale, Mike, "F-106B Delta Dart," Scale Aircraft Modeling, Vol. 5, No. 5, February 1983, Page 214

This article covers converting the Hasegawa 1/72nd scale F-106A to an F-106B. However, no mention is made to the change in the weapons bay. The scale drawings are claimed to be 1/72nd scale, but are too long in length, and too short in span. The bottom view is of an F-106A, not an F-106B as claimed. Again, the weapons bay is incorrect.